The term eco-system has over the last few years migrated from the world of environmental sciences to the world of manufacturing and industry. This is a rather strange journey and has been a source of concern amongst some scientists, engineers and business scholars, lacking as it does clear and precise definitions.

The advantages of an eco-system approach however are substantial, reflecting as it does the complexity and diversity of modern industrial systems. It is no longer sufficient to talk about subcontracting or even supply chains as it is now clear that many of the actors involved in bringing products to market may reside in different companies, different industries and indeed different countries. Capturing this complexity and the dynamic nature of modern industrial systems is vital if we are to understand how they work and how they might develop. This book is a timely and comprehensive attempt to put more structure into the discourse around business eco-systems. The authors have provided a good blend of theoretical development illustrated by practical examples of modern business eco-systems. Students, scholars and practitioners will find much here to interest them. – *Professor Sir Mike Gregory, Institute for Manufacturing, University of Cambridge.*

Business Ecosystems by Ke Rong and Yongjiang Shi is a landmark in the field of business strategy. As someone who has lived my life with managers developing business ecosystems, I can attest that the authors 'get' the essence and the power of the approach. Business ecosystems are the dominant design for strategy making in technology-based businesses today. In practice, business ecosystems are everywhere: producer-centered, customer-centered, people, technology and product centered. Business ecosystems nest within others. Business ecosystems are themselves complexly related.

The authors provide a model for studying business ecosystems in their richness. They review two decades of academic research in order to clarify the construct. The authors show that business ecosystems dynamics reflect the principles of general systems theory, agent-based modeling and the mathematics of networks. Helpfully, the authors demonstrate this by exploring the logical extension of leading systems-based concepts of advanced manufacturing into the domain of business ecosystems.

They demonstrate that the business ecosystem field of application is at a higher logical type than other theories of strategy – that is, business ecosystems ideas guide leaders to intervene to continually reshape industry structure, and to do so simultaneously within multiple related industries. Leaders collaborate to establish ecosystem-wide shared values and visions that in turn support collective conduct and result in shared gains in performance. Business ecosystems are notoriously difficult for outsiders to study. The guiding visions of business ecosystems are inherently cross-company and cross-industry, are usually held secret by members, and peer far into the future.

Ke Rong was able to gain access to top leaders in three related very large-scale global business ecosystems, originating on three different continents and in three forms of capitalism, all contributing to one of the most dynamic fields of world business. The result is a narrative of great interest to executives as well as researchers. By sketching the story in its broadest and most complete form, there is much for the rest of us to chew on, refine and question. The breakthrough is that we can do so as a community, with this work and its methodology as a foundation.

James F. Moore,
Concord, Massachusetts, December, 2014

Business Ecosystems

Constructs, Configurations, and the Nurturing Process

Ke Rong
Bournemouth University, UK

and

Yongjiang Shi
University of Cambridge, UK

First published 2015 by
PALGRAVE MACMILLAN

Palgrave Macmillan in the UK is an imprint of Macmillan Publishers Limited, registered in England, company number 785998, of Houndmills, Basingstoke, Hampshire RG21 6XS.

Palgrave Macmillan in the US is a division of St Martin's Press LLC, 175 Fifth Avenue, New York, NY 10010.

Palgrave Macmillan is the global academic imprint of the above companies and has companies and representatives throughout the world.

Palgrave® and Macmillan® are registered trademarks in the United States, the United Kingdom, Europe and other countries

ISBN: 978–1–137–40590–6

This book is printed on paper suitable for recycling and made from fully managed and sustained forest sources. Logging, pulping and manufacturing processes are expected to conform to the environmental regulations of the country of origin.

A catalogue record for this book is available from the British Library.

Library of Congress Cataloging-in-Publication Data

Rong, Ke, 1984–
    Business ecosystems : constructs, configurations, and the nurturing process / Ke Rong, Yongjiang Shi.
        pages cm
    ISBN 978–1–137–40590–6 (hardback)
    1. Industrial management. 2. System analysis. 3. Business planning. 4. New products. I. Shi, Yongjiang. II. Title.

HD31.R654 2014
658—dc23   2014025131

Contents

List of Figures

List of Tables

Acknowledgements

Many thanks go to the industrialists who provided access to their companies and allowed us to participate in the cases and research.

List of Abbreviations

API	application programming interface
BE	business ecosystem
BELC	business ecosystem life cycle
CISC	complex instruction set computing
CMOS	complementary metal oxide semiconductor
CP	content provider
DFM	design for manufacturing
EDA	electronic design assistant
GEN	global engineering network
GMVN	global manufacturing virtual network
IC	integrated circuit
IDH	independent design house
IDM	integrated device manufacturer
ILC	industry life cycle
IMN	international manufacturing network
IP	intellectual property
ISA	instruction set architecture/international strategic alliance
ISV	independent software vendor
MID	mobile internet device
MTK	Mediatek (Taiwan IC Company)
NPD	new product development
ODM	original design manufacturer
OEM	original equipment manufacturer
OMS	open mobile system
OS	operating system
OSV	operating system vendor
PC	personal computer
PCB	printed circuit board
PLC	product life cycle
RF	radio frequency
RISC	reduced instruction set computing
SC	supply chain
SI	system integrator
SN	supply network

TFT-LCD	thin film transistor liquid crystal display
VC	value chain
UMPC	ultra mobile personal computer

1
Introduction

1.1 New industry frontier: from supply chains towards business ecosystems

The mobile phone industry has experienced dramatic changes in the last five years. In the West, while Apple has dominated the industry and Samsung has risen in popularity, previously established players such as Nokia, Sony-Ericson and Motorola have almost disappeared. In the Chinese market, the new smartphone company Xiaomi sold almost 19 million smartphones in 2013, up from only 400,000 in 2011. By taking advantage of established manufacturing resources and integrating them on its business platform, Xiaomi successfully imitated Apple's business model, tailoring it to the Chinese mobile phone market. It co-opted the Chinese mobile phone ecosystem into its value-creation network, delivering unprecedented rapid growth and creating the most popular Chinese domestic mobile phone brand within two years.

Xiaomi's success may demonstrate the end of an era when the vertically integrated firm has been the dominant mode of industrial organisation in the emerging and fast-growing phases of industry development. Its alternative model illustrates that collaboration among complementary organisations, diversified resources and skill sets are essential for nurturing new firms, supply chains, value networks and industries.

A second example of rapid change is that the mobile phone industry has also converged with the personal computer (PC) industry to form a new mobile computing industry, potentially improving the performance for portable devices. Two groups of companies have been engaged in developing potential end-user products in this emerging industry. The mobile phone group produces a portable device used for daily communication while the PC group is normally involved in data processing or entertainment. However, consumers have an increased expectation to be able to carry out various simple computing operations while they travel, and mobile computing functions such as easy access to the Internet, long standby time and simple computing have become very appealing. To meet these expectations, the mobile phone group has devoted more attention to smartphones or MIDs (Mobile Internet Devices) than to the 2G phone in order to add more computing functions to enable users to carry out simple computing tasks while travelling, while the PC group seeks to make the notebooks much smaller in size, portable and with long standby times, so that the product can perform more functions without being recharged.

However, the convergence of these two industries is not straight-forward. Products of the mobile computing industry are still in flux, although iPad dominates the market. Each group hopes to retain its advantages to form new generations of products. They cannot reach an agreement in the design of end-user products. More critically and inter-estingly, the industrial structures and the business models in the two groups are very different. The PC industry has been dominated by the Win-Tel system for more than 30 years while the mobile phone industry is mainly enabled by ARM IP architecture and more diversified networks domination. The serious competition between these two groups inspires and forces many firms to concern themselves deeply with their selec-tion of business ecosystems (Moore 1993). The new competition in the emerging industry has been transformed from the traditional levels between firms and supply chains towards a new level between Win-Tel–based and ARM-based ecosystems.

The third example has nothing to do with mobile phones but goes back to China again. The Chinese central government has invested a huge amount of money in renewable energy vehicles in the last fifteen years (from its 9th to the 12th Five-Year-Plan periods). Strategically it intends to find new ways to develop the family transportation industry and to solve two longer-term issues. First, although China has emerged as the largest producer and market in the automotive industry, the Chinese government has been convinced that its state-owned car producers have become Chinese subsidiaries of foreign, multinational corporations (MNCs). They are weak in innovation and high-value–capturing capabilities. Secondly, the fusel fuel combustion engine automotive industry is almost towards its own end, and China has yet to develop its own renewable energy vehicle industry. The politically motivated investments have, unsurprisingly, not brought on any significant industrial breakthroughs. However, the grass-roots entrepreneurs, aligning with their local governments in the Shandong Province (located between Beijing and Shanghai), amazingly created a very successful but quite low-end and largely 'illegal' electric vehicle (EV) industry (it is also called micro-EV) in the last ten years. When its market size almost reached 200,000 units in provincial annual EV sales in 2010, the Chinese central government forbade the EV productions based on Chinese established regulations and policies in the automotive industry.

The serious arguments and thoughts are that, although there are very strong and increasing demands for the low-end EVs from the rural market and even stronger supply capabilities from the province and global industry supports, it is useless unless the automotive industrial policies, regulations and legislation are able to be adapted according to the new development requirements. The emerging EV industry must be the best example to demonstrate the increasingly critical challenge for an industry development. For the Chinese grass-roots EV manufacturing companies and their local and global collaborators, it is not the most serious challenge to understand the increasing demands for the EV, or to set up a supply chain or value network to co-develop the EVs to satisfy the targeted markets, or even to identify the complementors, such as electric charging service providers and their underlying renewable energy infrastructures. It is so essential that all parts – demand, supply, intermediary sides – work together. In the Chinese Shandong EV industrial development case, the intermediary part is obviously the most significant bottleneck.

As a result of the above three examples, industries are facing emerging challenges in order to cope with the dynamic changes and uncertain business environments as well as the fast emergence and transformations of new technologies and market demands. These challenges may not come mainly from an individual firm or supply-chain levels but from a more complex, dynamic and much wider range of business contexts and systems together. Industrial people adopted ecological metaphors and gave the new challenging totality a very imaginable terminology: business ecosystem (BE).

There is no doubt about the BE existence and its strong impact on industrial development and global competition. However, industry also asks, if the business ecosystem is so powerful and critical,

- what a business ecosystem is, and what key building blocks organise a business ecosystem;
- and how a business ecosystem should be nurtured.

1.2   Business ecosystems: from a fantastic metaphor to scientific understanding

Although academics have paid close attention to business environmental changes and features for a very long time, the concept of a 'business ecosystem' was only formally proposed in 1993. James Moore developed a theory to explore and explain the interactions and co-evolutions between firms and their business environment (1993). He defined the business-related firms and their environments from a totality named a business ecosystem. A business ecosystem contains opportunity space, supply-chain–related companies and different levels of organisations, such as industry associations, competitors and policymakers. These organisations have a big impact on the development of an industry, especially when that industry is experiencing rapid change. A business ecosystem has a life cycle of four sequential phases: birth, expansion, authorities and renewal. Moore's theory inspired firms with the importance of co-evolution, highlighting cross-industry activity and blurring industry boundaries (Moore 1996).

After Moore's paper, about 350 academic papers have been published in the last 20 years. So far the main contributions of business ecosystem literatures were from five bodies of knowledge, including strategic management, organisation and network studies, operations management, digital industry and, most recently, innovation management. These academic disciplines have taken business ecosystem as a unique lens through which to review their established boundaries and to identify emerging issues. It is obvious that many disciplines have realised that the new concept not only brings an attractive metaphor but also opens a much wider territory for exploration.

However, the research and related theories specifically targeting business ecosystems and their behaviours are still very rare, especially lack of systematic observation and analysis (Peltoniemi & Vuori 2004). In the academic world there seems to be still no consensus towards recognising the business ecosystem as an emerging research arena that can be fundamentally helpful to business and management.

Currently the industrial challenges from many emerging industries such as the Internet, mobile computing, renewable energy and 3D printing, as well as the digital economies with uncertain product and complex collaboration, have highlighted the importance of business ecosystem utilisation. This phenomenon indicates a good opportunity to study the business ecosystem more deeply.

Business ecosystem of course can be an excellent metaphor to help the imagination of the business environment, targeting its latent resources, leveraging a better position strategically and forming a unique value-creation network more effectively.

However, it is much more critical for academics to realise the business ecosystem as an emerging and essential body of knowledge and a rich research arena. Therefore it is worth thoroughly exploring and understanding the attributes and behaviours of business ecosystems.

1.3   Aims and characteristics of the book: explorations of the system and methodology

This book seeks to explore the basic characteristics of business ecosystems and to capture some practical experiences to nurture an ecosystem for business transformation. In order to comprehensively study business ecosystems, a systematic investigation has been conducted to build the theories relevant to the business ecosystem's concerns.

During globalisation, the manufacturing industry has been fundamentally transformed from the traditional, single-site factory- or firm-based system into multiple levels of networks, including intra-firm (international manufacturing) networks, inter-firm (supply chain, industry cluster, business) relationships, as well as the international and inter-firm (international alliance, global manufacturing virtual network, open innovation) networks (Shi & Gregory 2001). The globalised transformations also change the business ecosystems. Evolutionarily the business ecosystem can also be recognised as a further extension of globalisation and transformation. It is a critical part of studying business ecosystem to deeply understand the transformation process, especially to capture the journey of emerging industry along its whole life cycle.

By learning from the research background, the main research objectives are to understand the business ecosystem as a whole, and to specifically:

1) identify key constructive elements of a business ecosystem
2) explore the evolutionary path of a business ecosystem
3) summarise typical patterns of a business ecosystem with adaptive strategies
4) develop a firm-based business ecosystem nurturing process

1.4   Structure of the book chapters: a research journey

As shown in Figure 1.1, this book consists of three main parts and 12 chapters.

Chapter 1 introduces the background and an overview of the book.

Following the research journey, the main body of the book is comprised of three sections between the introduction and the conclusion:

- Part I: Background Exploration of Business Ecosystems: In this section, industrial, practical and academic literature reviews have been conducted to discover research gaps and to inform the research framework and process. It lays down a theoretical foundation and sets up research directions on business ecosystem theories.
- Part II: Case Observation of Business Ecosystems: This section, consisting of three chapters, gives detailed industrial stories from three typical industrial companies: ARM, Intel and MTK. In each main case study there are also three sub-cases to inform understanding of the business ecosystem nurturing.
- Part III: Theory Construction of Business Ecosystems: This section includes four chapters dedicated to four key research findings and their discussions on business ecosystem theories: first, the life-cycle phases are developed to describe the ecosystem's growth; second, constructive elements are identified based on the 'structure–infrastructure' model; third, the different configuration patterns are defined according to the selected dimensions; and, finally, the firm-based business ecosystem nurturing process was conceptualised and developed for further verification.

The final chapter is the conclusion which encapsulates all the content of this book as well as offering suggestions for future research directions.

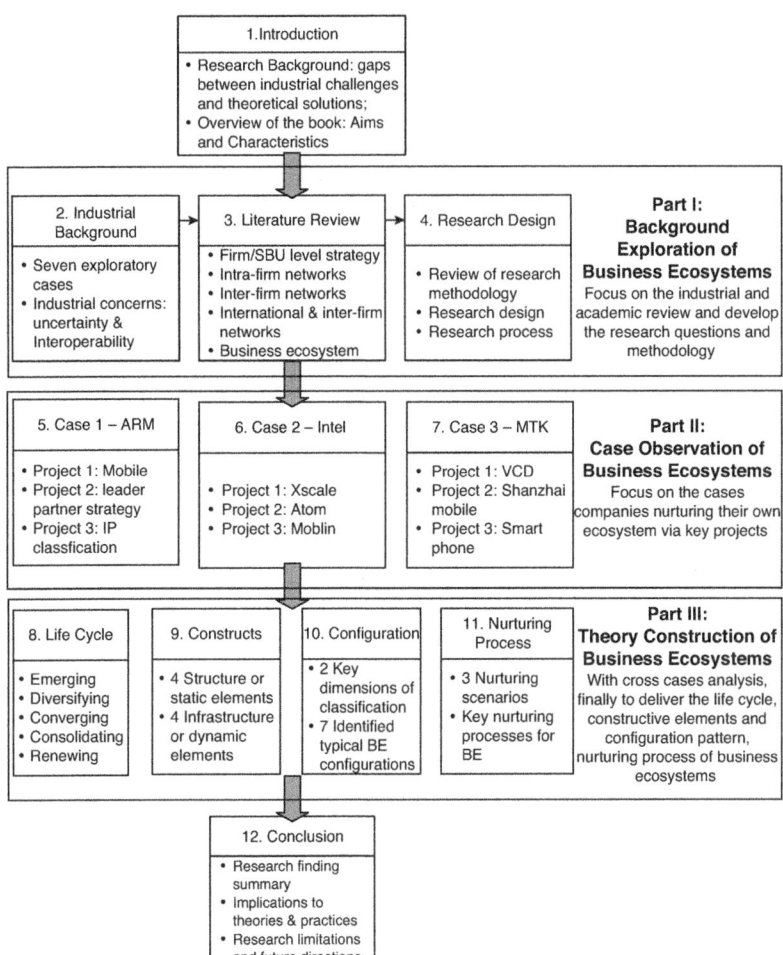

Figure 1.1 Section and chapter structure of the book

Part I

Background Exploration of Business Ecosystem

Part I focuses on the reason why business ecosystem research is required from both the industrial and academic perspectives. This part also develops the research questions and methodology to prepare the further exploration of the business ecosystem theories,

2
Industrial Challenges

2.1 Introduction

By studying exploratory cases in the mobile computing industry, this chapter seeks to explore and discuss emerging industrial issues with which the business ecosystem research is booming.

The objectives of this chapter are:

1) to introduce the evolutionary path of the semiconductor industry in order to explain the background of the emerging mobile computing industry;
2) to explore the challenges facing the mobile computing industry as it emerges from the convergence of the mobile phone and PC industry;
3) to deeply analyse the industrially challenging issues of uncertainty and interoperability as the reason for business ecosystem research.

2.2 Industry background review

2.2.1 Semiconductor industry brief review

Firms in the semiconductor industry design Integrated Circuit (IC) chips, manufacture them from wafer, and also sell them to Original Equipment Manufacturers (OEM) for use in applications. Figure 2.1 shows the general areas (design, manufacturing, application) of the semiconductor supply chain. The following section will address the evolutionary stages of semiconductor industry development, its industry structure specialisation process, as well as its technology architecture choices.

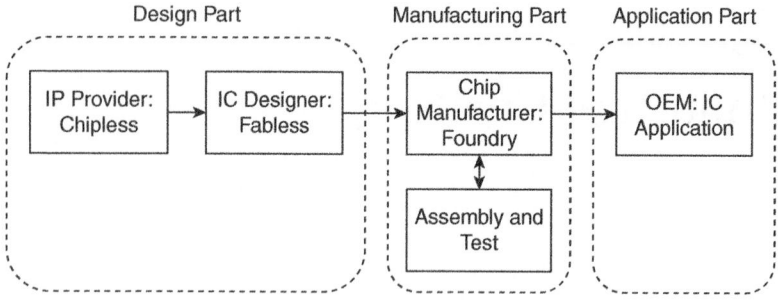

Figure 2.1   Semiconductor supply chain

1) Five-stage evolution of the semiconductor industry

The semiconductor industry has evolved through five stages: introduction, Integrated Device Manufacturer (IDM) dominating, foundry
booming, IP firm emerging and Renewal as shown in Figure 2.2 (Rong
et al. 2008). In the introduction stage, the semiconductor industry origi-
nated in the invention of transistors in 1947 at the Bell Labs in the
United States. American firms dominated the industry at the beginning, largely due to strong government support for national defence
purposes.

In the IDM-dominating stages before 1990s, the most prevalent business model was to adopt the IDM model, which was to include all three
areas of the semiconductor supply chain: IC design, manufacturing and
application as shown in Figure 2.1. Normally the big firms like Motorola
and Intel were regarded as IDM who kept all three areas of the supply
chain within their own firms. At this stage US firms kept their advantage by designing and introducing the mass production technology of
complementary metal oxide semiconductor (CMOS) in 1963 and its first
significant application, the Palm calculator, in 1967. However, in the
1980s Japanese firms gradually built up their leading positions ahead
of US firms through establishing strategic alliances with the US firms,
acquisitions and self-development (Borrus et al. 1986). Subsequently,
US firms had to work with Taiwanese and Korean firms to recapture
the market competitive advantage, which triggered the third stage, as
'foundry booming'.

In the third, foundry-booming stage, the semiconductor industry experienced a huge change in its structure. In the 1990s the 'pure foundry'
companies were developed in Taiwan (Shih & Wang 2009), and these

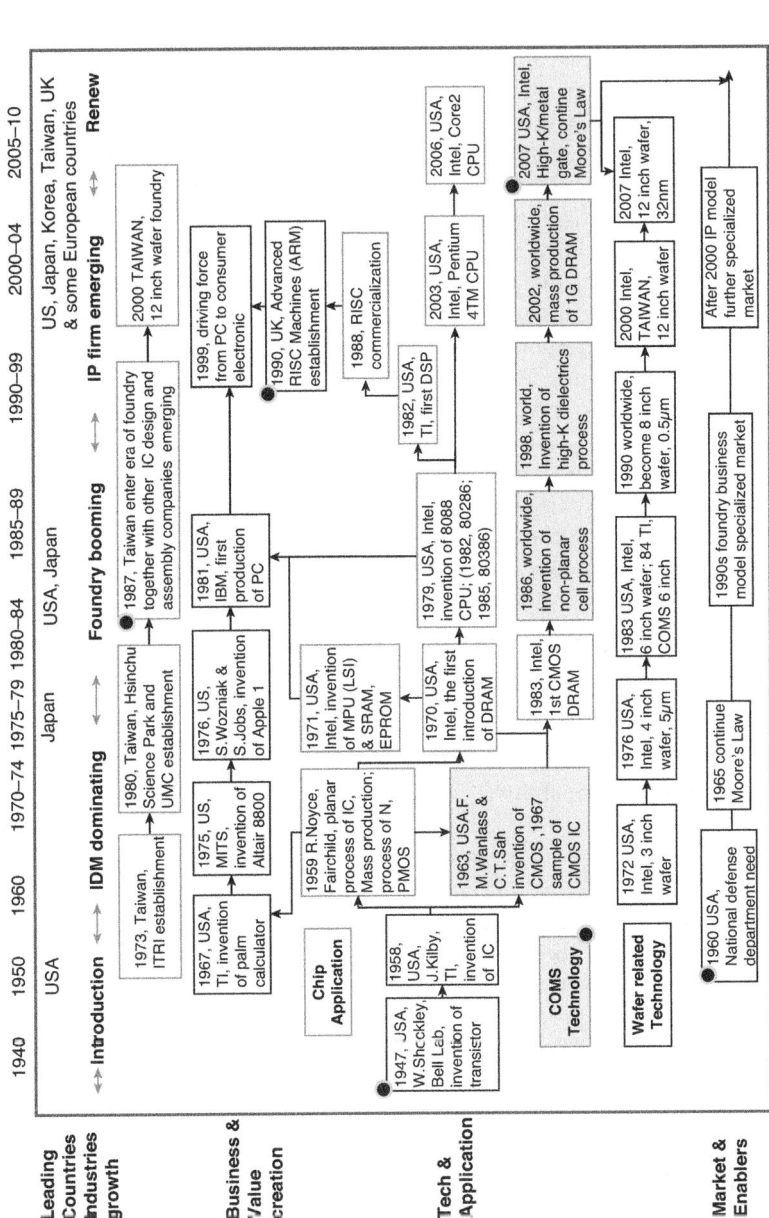

Figure 2.2 Roadmap of semiconductor industry

provided the chips manufacturing service for IC design companies. This in turn stimulated the growth of many IC design companies. IC design companies were less capital intensive than the IDM companies, and this lowered the entry barrier for many latecomers. The existence of foundry companies allowed the semiconductor supply chain areas to become specialised and fragmented. The two different business models of IDM and horizontal coordination could coexist harmoniously.

In the fourth stage, as IP (Intellectual Property) firms emerged, the most popular business model was the IP license model. In light of the phenomenon that many IC design companies spent a huge amount of money in designing chips with the same functions as other companies' chips, some responded by focusing more on the design of specific functions, packaging them into IPs and then licensing them to IC design companies. A typical example was ARM, one of the strongest IP providers in this industry. It provided the IP that is regarded as the basic architecture of the IC chip. As a result, IC design companies started to design chips by combining different IPs and their own designs to deliver chips for different markets, including the mobile, computing, embedded and home market. ARM charged a license fee and royalty fee on the new chips.[1] This IP business model has triggered an increasing specialisation and fragmentation of the semiconductor supply chain since the 1990s.

In the fifth, 'renew' stage, as the semiconductor industry faced the limitations of manufacturing technique, companies endeavoured to introduce more significant technology, and also built alliances to strengthen their capability.

2) Two technology architectures for core processors

The term 'chip' or core processor is used here as a shorthand and generalisation for 'the core data handling component which enables the function of any electronic device'. There are two main technology architectures for core processor chips: reduced instruction set computing (RISC, like ARM's architecture) and complex instruction set computing (CISC, like Intel's architecture-X86) as shown in Figure 2.2.

Architecture is often interpreted with two separated meanings: ISA (Instruction Set Architecture) and micro-architecture. ISA is a part of the computer architecture related to programming, including the native data types, instructions, registers, addressing modes, memory architecture, interrupt and exception handling, and external I/O. Micro-architecture is the way that a given ISA is implemented into a processor (IP).[2]

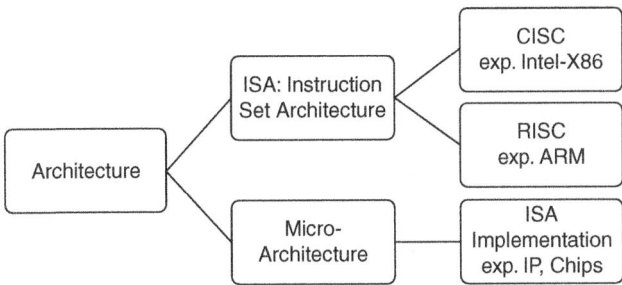

Figure 2.3 Two architecture choices for chip design

Figure 2.2 uses leading firms (ARM and Intel) to explain the difference between the two architectures: ARM has grown from the mobile phone (embedded) market with low power, small-sized and highly integrated architecture, while Intel has grown from the PC industry with the architecture of a strong computing capability. Briefly, CISC architecture is often used as the PC's processor for work involving heavy-duty computing, and RISC is used more in the 'embedded' market: home devices, mobile phones and for portable and low-level computing use.

Recently, these two markets (embedded and PC industry) have begun to converge, and this has resulted in competition between the two groups, who used different technology architectures.

2.3 The mobile computing industry brief review

2.3.1 General evolution: convergence of the mobile and PC industry

The rapidly expanding technology of cellular communication, wireless connection and GPS service will make information accessible anywhere and at any time. In the near future, tens of millions of people will carry portable devices to connect with the network instead of accessing from a fixed physical position (Imielinski & Korth 1996). Mobility and portability will create an entirely new class of applications and new massive markets combining personal computing and consumer electronics. From the industry perspective, practitioners such as Intel, ARM and MTK regard the mobile computing industry as the convergence of the mobile phone industry and PC industry. On the application side, the mobile computing industry is stimulated to develop many product designs to meet the variety of the massive markets.

As shown in Figure 2.4, the mobile computing industry emerged as the convergence progress accelerated between the mobile industry and PC industry. The picture demonstrates the convergence trend by using two dimensions of function and portability. In terms of functionality, customers not only want to use the phone function, but also want to implement mobile computing service including email, video and entertainment. In the dimension of portability, the laptop as the main product of the computing industry is not portable everywhere and does not have easy access to the Internet. So ideally this desirable device should be enabled to be always on, always available and always connected to the Internet (Kenney & Pon 2011; Rong, Shi, et al. 2013). However, so far the concept named MID (Mobile Internet Device) has not yet been transferred to a best solution. All the companies from different backgrounds are working together to propose the final solution for this concept. Although they have managed to come up with a few devices that possess the MID characteristics, no firm has managed to formulate an ideal product fully satisfying the customer's requirement.

The device comprises many parts, including hardware: the baseband processor, application processor, power management and RF (Radio Frequency) processor; software: operating system, application software and much embedded software for communication. Therefore various companies will be involved in developing the MID as Figure 2.4

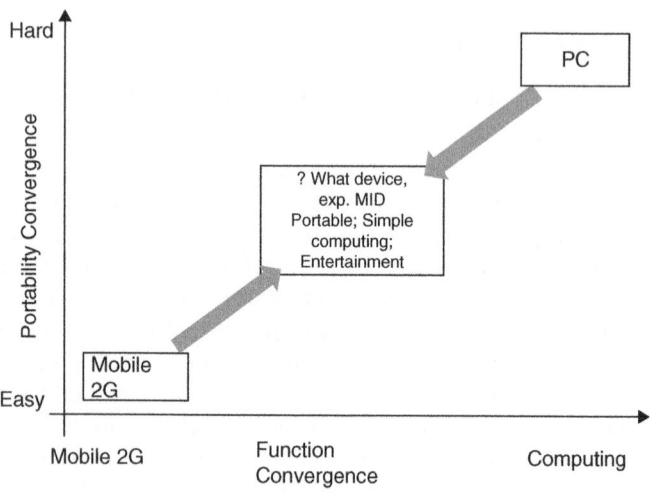

Figure 2.4   The convergence trend in the mobile computing industry

2.3.2 High interaction in the mobile computing industry network

Figure 2.5 was developed from exploratory cases and the semiconductor industry supply chain. It demonstrates the key players' network in the mobile computing industry. In order to describe the industry unequivocally, abbreviations are listed and explained first in Table 2.1. Generally, each practitioner regards its product as a platform that enables partners to interact more easily and efficiently. For example, OSV regards its operating system as a platform where ISV can develop application software.

Table 2.1 Players in the mobile computing industry

Players	Definition and function	Challenges
OSV: Operating System Vendor	OSV is short for operating system vendor, who produces the operating system for MID.	Various operating systems, not compatible with one another
ISV: Independent Software Vendor	ISV is short for independent software vendor who designs the application software running on the specific operating system.	The software products have to be compatible with various operating systems
CP: Content Provider	CP is short for Content Provider who provides the service via the different application software.	Have to be compatible with different application software
Foundry	Foundry companies manufacture chips after design companies' design.	Manufacture different technology architecture
IC Design: Integrated Circuit Design	IC design companies design different chips with specific functions.	Selection of different IPs; attract more OSV support and OEM adoption
EDA: Electronic Design Automation	EDA is short for electronic design automation which provides the software for designing electronic systems like integrated circuit chips.	Persuade IP and IC firms to adopt its design tools
IP providers: Intellectual Property Providers	IP provider is short for intellectual property providers who provide IP as the basic part of chips design which could be repeatedly used and have different functions.	Persuade IC design firm and foundry to adopt its technology architecture

Continued

Table 2.1 Continued

Players	Definition and function	Challenges
OEM: Original Equipment Manufacturer	OEM is short for original equipment manufacturer who is the brand owner and coordinates the network partners.	To coordinate different components along the supply chain
ODM: Original Design Manufacturer	ODM is short for original design manufacturer who delivers total solution designs for OEM's new products.	To persuade OEM to adopt its total solution
EMS: Electronics Manufacturing Service	EMS is short for electronics manufacturing service which provides the design, manufacturing, purchase, logistics and assembly of electronics products for OEMs.	To persuade OEM to outsource their manufacturing
Standards organisation	Standards organisation sets rules that OEM and carrier should adopt.	To coordinate industrial players
Carriers	Carriers are the network operators who provide the infrastructure that enables mobile computing devices to connect to the Internet.	Not only a tube owner but may penetrate into other service areas

Also, this kind of operating system is compatible with IC firms' chip platform. In general, the end-user products are very complicated: every component should be compatible with the others, which requires a high level of interaction between partners.

Figure 2.5. shows the high level of interaction within the mobile computing supply chain, which is a specialised subset of the semiconductor industry of Figure 2.1, and also adds in more functional players from the application side. Normally, OEMs need to obtain the application processor and baseband processor from the **design part** such as the IC design companies who also license IP from IP providers. EDA software companies provide the software designing processor to IP provider, IC firms and foundry players.

In the **application part**, more firms emerge with different functions: in order to meet the customer's requirement, the device needs operating systems from OSV and application software from ISV. This

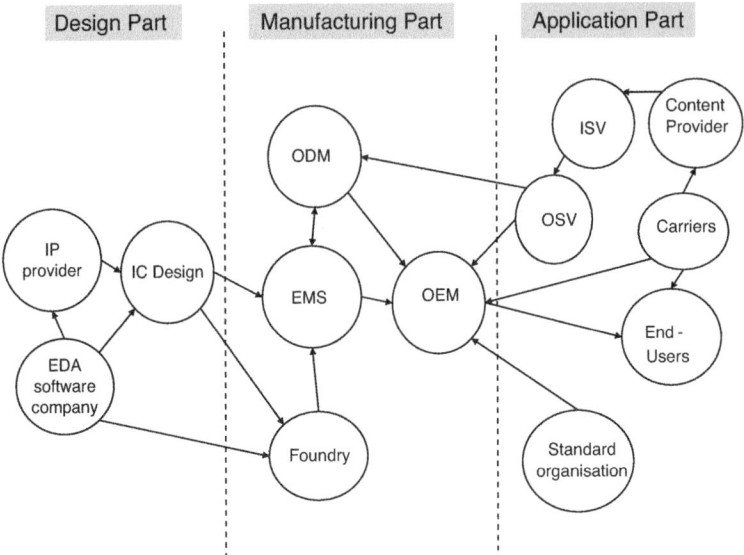

Design Part   Manufacturing Part   Application Part

Figure 2.5 Key players in the mobile computing industry
Source: Adapted from Rong et al. (2013).

device is operated by some carriers under certain standards. The content provider provides content based on various specific application software. ODMs usually help to offer the total solution of the final device for OEM. Standards organisations help coordinate partners and nurture future industry standards. In the **manufacturing part**, foundry companies help manufacture chips which are similar to that of the semiconductor industry. In conclusion, in the mobile computing industry, no single company can approach the final device. Instead, it needs companies to network and to work collaboratively to propose future products.

2.3.3 Uncertainty about end-user products

So far the market is still highly uncertain. There are eight kinds of devices listed in the mobile computing industry which still have not met the requirements of customers and industrialists in several aspects, including portability (screen size), function and working hours in Table 2.2.

As the market is very fragmented, the dominant design has not been finally confirmed. It makes companies within this industry collaborate with one another to a high degree in order to accomplish the best

Table 2.2 List of current solutions for the mobile computing industry

Property	Smartphone	UMPC	MID	Smart-Book	iPad	Netbook	Ultrabook	Notebook
Processor (normally)	ARM	ARM	ARM Intel	ARM	ARM	Intel	Intel	Intel
Display (inch)	2-3	3-5	5-7	7-10	9.7	10-12	13-15	13-17
Connectivity	3G, 2G, Bluetooth	WiFi, Bluetooth	WiFi, Bluetooth	3G, WiFi, Bluetooth	3G, WiFi, Bluetooth	WiFi Bluetooth	WiFi Bluetooth	WiFi Bluetooth
Interface	Camera, audio	Audio	Camera, audio USB	Camera, audio USB	Camera, audio	Camera, audio USB	Camera, audio, USB	Camera, audio USB
Weight	200-300g		<850g	800g	469-662g	800-1.3kg	<1.5kg	>1.5Kg
OS	Symbian, Windows mobile Android	Linux	Linux-based Android	Linux-based Android	iOS	Windows, Linux-based	Windows	Windows
Battery life (hours)	24	3	9	9	10	5	>5	3
Price dollar (normal)	200-500	N/A	700	200-300	399-929	250-399	650-1800	400-3000

solution as soon as possible. Companies start not only focusing on their own parts of the business, but also look at the whole industry and encourage other partners to contribute.

2.4   Exploratory case studies: key players in the mobile computing industry

The following seven exploratory cases were conducted with both semi-structured interview and second-hand data collection by the researcher since 2009 in China. These seven cases are chosen as the typical players with different complementary positions in the mobile computing industry, which together form most parts of the value chain as Figure 2.6 (Rong, Shi, et al. 2013). Furthermore, these seven typical cases highlight the outstanding industrial challenges and also introduce the companies' strategies and actions to overcome those difficulties.

The exploratory case studies are structured with the following steps: 1) brief introduction to the company; 2) industrial challenges; 3) the way to deal with those challenges; 4) the results for the case company; 5) learning points from this study.

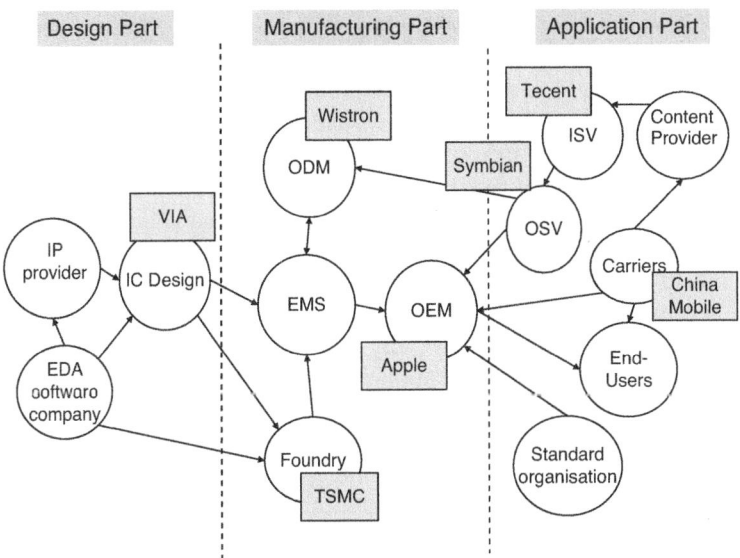

Figure 2.6   Exploratory cases' position in a typical mobile computing supply chain

2.4.1   Symbian: operating system vendor (OSV)

Symbian, as a joint venture, has been an operating system vendor in the mobile phone industry since 1998. An operating system (OS) is software that acts as an intermediary between a user of a computer and the computer hardware (Silberschatz et al. 1991), that manages the device's hardware resources, and that also provides service as a platform for the execution of application software (Wood et al. 2005). Various operating systems from competitors emerged due to the huge increase in the mobile phone market. In order to maintain its competitive advantage, Symbian has kept in close cooperation with the OEMs and encouraged independent software vendors (ISV) to develop applications based on Symbian's OS. Symbian successfully enabled the explosion of mobile device innovation, and achieved sales of 100 million mobile handsets by 2006.

However, in 2008 Smartphone OS became popular as the mobile computing industry emerged. Symbian began to lose their market share as their OS mainly focused on the feature mobile phone (2G). Smartphone OS enabled many novel functions like entertainment, online camera, simple computing and Internet access which greatly extended the functional scope of the mobile 2G. These requirements also triggered the close collaboration between the OSV and ISV. In the meantime the Smartphone OS was not finalised well. The competition for industry standard came up as various OSVs penetrated into this emerging market. As the market requirements were very dynamic and uncertain, different OSVs had to persuade OEMs to adopt their OS, and they were also more open to ISVs. So, in order to encourage those partners to nurture the future products based on Symbian OS, one shareholder – Nokia – purchased all Symbian's assets and turned Symbian into an open source – Symbian Foundation. Nokia lost the benefit from Symbian OS, yet they got benefit from the Smartphone development project in return because, after the establishment of Symbian Foundation, firms were encouraged to use any source of Symbian Foundation to propose new OS and relevant application software. So these new OS and software supported Nokia's Smartphone development and made Nokia Smartphone diversified and innovative, which enabled Nokia phones to meet uncertain business requirements.

Symbian provided a good example of the adaptation to the uncertain business environment: 1) first, Symbian recognised the uncertain environment of mobile computing; 2) Symbian then became open to partners, and initiated the open source to highlight collaboration; 3) open source foundation not only encouraged innovative and flexible collaboration; but 4) also assisted firms to achieve product variety and 5) shortened the products' lead time for market quick response.

2.4.2   Tecent: independent software vendor (ISV)

Tecent, as an ISV, is providing an online messenger. QQ – which has already profoundly influenced people's virtual and real communication in China.

As a supplementary role in the mobile computing industry, Tecent also faced challenges, including 1) meeting the requirement of customers' online communicating and entertainment; 2) compatibility and embeddedness to different operating systems. There were many OSs (operating systems) in the mobile computing world. And every OS needed Tecent's customised instant messenger application software.

Tecent tackled these problems in two ways. First, Tecent paid much attention to maintaining the relationship with many OSVs. They delivered specific versions of instant messengers to different OSs and kept updating them. Tecent not only followed OSV's instructions, but also set up specific alliances to maintain their tools' compatibility and ease of embedding in various OSs. As a result, Tecent developed more than 20 versions of instant messengers compatible with different OSs in order to cope with customers' specific requirements.

Secondly, Tecent aimed at building up the online community that presented most of people's online virtual life network as well as their real life network. This community included instant online messengers, information websites, search engines, online games, e-business and personal Internet space with virtual products. The number of registered instant messenger customers had already reached 990 million by 2009. This in return also drove many partners to improve their products to be compatible with Tecent's software.

Some good points could be learned from Tecent: 1) they closely followed the OSVs' instructions and maintained the relationship with them;. 2) they formed some specific alliances to facilitate their instant messenger in becoming more generalised and well accepted to reduce the uncertainty; 3) they expanded their own business scope by building up an online community platform which combined people's real life and online virtual life.

2.4.3   TSMC: foundry – more than a pure foundry

TSMC was established in 1987. It has developed as the largest dedicated semiconductor foundry in the world and has consistently offered the foundry segment's leading technologies (32nm) and design services to its customers. In 2007 its total revenue represented 50 per cent of the dedicated foundry segment within the semiconductor industry.

However, TSMC has faced many uncertainty challenges due to the nature of manufacturing: 1) uncertain orders: foundry companies cannot predict the amount of orders. The overhead costs of foundry companies is extremely high, so the machinery must work as continuously as possible; 2) manufacturing feasibility: IC design firms designed their chips based on different technology architectures. IC design companies hoped to test their chips for manufacturing feasibility in advance; 3) IP compatibility with design tool firms, IP provider and IC design firms.

In order to deal with these uncertainties, TSMC adopted the strategy of DFM – design for manufacturing ecosystem. In this ecosystem, TSMC would share IPs of specific functions with design tool companies, IP providers and IC design firms and finally formulate an IP library. The shared IPs would help design companies fast design chips with manufacturing feasibility. Furthermore, this ecosystem also encouraged every partner inside to share its own IPs in advance to test the manufacturing feasibility, so that the manufacturing process could be straightforward with short lead time and low cost. If this ecosystem became stronger, it would attract and deliver more offers to the new participator.

From TSMC's case, some conclusions can be made: 1) TSMC dealt with very uncertain market demand; 2) they set up the DFM ecosystem to extend their partners' scope from IC design company towards design tool company and IP providers; 3) partners inside this ecosystem were very flexible to get IPs. They also contributed IPs to enrich the database; 4) thus it made TSMC ready to manufacture and achieve quick response; 5) it attracted more customers to cut down the cost.

2.4.4   China Mobile: carrier

China Mobile was one of the largest telecommunication carriers, having as many as 500 million registered customers. Due to the mobile computing industry emerging, the related services were regarded as a more significant area for growth than to simply remain a 'tube lord' (i.e., one who provides the connectivity conduit). Customers began to use mobile phones as one of the key entertainment tools to access the Internet, play games and communicate with friends.

In order to appropriate more value from the mobile computing industry, China Mobile expanded its business scope and launched an operating system project – OMS (open mobile system). This OS was developed from Android, which was open-source software developed by Google. Thus it could be compatible with many devices already listed in the market. This project was to encourage partners to develop application software based on their own operating system. China Mobile could encourage other ISVs to develop applications based on this OS and also embed more service into this OS. As a result, China Mobile phone could charge many fees from the services and applications for customers. In summary, China Mobile attempted to occupy the market for the operating system and services level instead of being just a 'tube for communication'.

China Mobile set up a project specifically to expand their original activities: first, China Mobile licensed its 3G technology to IC design companies and persuaded many OEMs to adopt the chips and their own Oss; second, China Mobile supported ISVs with a software development kit to encourage them to develop applications based on their OS. These applications were finally embedded into the new mobile phone. Many of these applications required a service fee from customers; third, when the phone came to market, China Mobile would do some collaborative sales activities to promote the new phone.

From China Mobile's strategies, there were also some learning points: 1) the 3G era required more services, rather than simply the mobile itself, and this resulted in many uncertain market requirements; 2) China Mobile shared its resources and enabled partners' collaboration; 3) China Mobile built up the OS ecosystem and formulated a very flexible network; 4) China Mobile provided more services to customers, which increased the variety of products for the OS.

2.4.5 VIA: IC design firm

VIA was the chip processor manufacturer. It was so great an opportunity for VIA that the mobile computing market developed with a huge growing demand for processors. The Netbook – a low level of notebook as the key representative product – has been extremely popular in the mobile computing industry since 2008.

However, VIA came across some challenges: 1) Intel, due to its strong experience in the computing industry, was soon securing competitive advantage in terms of Netbook market; 2) the standard of Netbook's product specification was not finalised, and VIA regarded Netbook as an opportunity to compete with Intel; 3) Netbook also could not meet the uncertain market requirement of the mobile computing industry.

In order to deal with the above challenges, VIA launched an alliance to encourage the partners to work together based on its chip solution. This alliance invited all types of players in the mobile computing industry, including software and hardware suppliers and OEMs. These players would finally formulate an integrated supply chain for Netbook production based on VIA's chips' solution. By 2009, around 47 companies had become members of this alliance. Furthermore, this alliance was open to all OEMs who wanted to deliver Netbook or other devices in the mobile computing industry. With the efforts from alliance members, the end-user products could be diversified and delivered in a short time to cope with dynamic market requirements.

This alliance played a significant role in new product development: 1) it was an integrated supply chain where every player would contribute its parts to spur various innovative activities; 2) it then accelerated innovation to achieve better solutions in this emerging and uncertain market; 3) it also reduced the industry entry barrier in the mobile computing industry, and encouraged small players to contribute their specific parts; 4) it provided the turnkey solution (total solution) for mobile devices by formulating an integrated supply chain which would enable the downstream partners' innovation and reduce the cost; and 5) it shortened the delivery time for new products.

2.4.6   Wistron: Original Design Manufacturer (ODM)

Wistron was one of the largest ODMs producing ICT (information and communication technologies) products. With its strong engineering background, Wistron was able to help their customers with new product development.

As an ODM, normally, Wistron offered a 'total solution' package for some products to OEMs who would feed back information. Recently, as the mobile computing industry emerged rapidly and the end-user product was not finalised, Wistron not only played a vital role in coordinating with original suppliers to integrate a new solution for this industry but also provided total solutions to PC OEMs, telecom carriers and mobile OEMs. This is because they all wanted to penetrate the mobile computing industry via different routes.

Wistron aimed to design well-accepted products by acting on feedback from three types of customers (PC OEM, telecom carrier and mobile OEM). As shown in Table 2.3, Wistron launched a P/Book project which was an improved Netbook as a potential solution for the mobile computing industry. This project would demonstrate a high degree of partner collaboration inside this industry. The P/Book was designed for two purposes: 1) entertainment – in order to provide a better experience of watching movies P/Books had a wide-size screen (1280*545); 2) business – P/BOOK products could display two windows in the screen, allowing users to work and have entertainment at the same time.

Table 2.3 An example of P/Book product detail

Collaborating partners	Detail	Industry position
TSMC	Wafer manufacturing	Foundry
ASE	Assembly and test	Assembly and Test
Qualcomm	Qualcomm QSD8250: double core: application processor and baseband processor	IC design
Redflag	OS support: Linux, Windows, Android	OSV
ANCO et.ac	Case	PC part
Compeq et.ac	PCB	PCB
	TFT-LCD 11.1inch, 1280*545	Display

Wistron was a good example to demonstrate how to capture new business opportunities by coordinating business partners: 1) Wistron delivered a novel product design after they identified the key requirements for the emerging industry; 2) they delivered products' solution for a wide range of customers, including mobile OEM, PC OEM and carriers; 3) they established solution platforms to cope with various requirements and so as to achieve a quick response to the market.

2.4.7 Apple: Original Equipment Manufacturer (OEM)

Apple, incorporated in 1977, is famous for its consumer electronics and software products, including computers, media devices and, later, mobile phones, as well as their online store (www.apple.com). Apple is in a leading position with the most fashionable product design and friendly user interface.

The services element was regarded as a more significant growth area in the smartphone industry in comparison to the 2G mobile phone industry. Apple was facing huge challenges against the traditional strong players like Nokia and Motorola. Apple introduced the iPhone (a smartphone) and built up online stores to solve those challenges. This smartphone combined the functions of the phone, media player with online store and Internet experience into a single, fashionable, easy-to-use device, which was more valuable to customers than other kinds of smartphones, and became their top choice.[3]

Customers registered in the Apple online store could download applications from this store sometimes free of charge. Besides, customers were encouraged to develop applications based on the iOS operating system while the API (application programming interface) was provided free by Apple. In this way more and more customers and small companies could earn money from helping Apple to deliver applications through the API. Meanwhile, the scope of the applications in the store was enriched hugely by the interactions among Apple, its partners and customers. This kind of ecosystem allowed the achievement of triple-win business.

Apple's online store as an ecosystem offered some learning points: 1) Apple shared resources such as the API to encourage partners' contributions to cope with market requirements; 2) Apple maintained the robustness of the community to attract more partners by setting up the working rules; 3) partners could deliver various value-added products and then they shared the value.

2.5   Key industrial challenges: uncertainty

From the exploratory cases, the industry uncertainty (Boretos 2007; Kenney & Pon 2011) is the most frequently highlighted industrial issue. As the mobile computing industry formed from the convergence of the mobile and PC industries, there were no well-accepted product designs. It is unequivocal that uncertainty is one of the key environmental forces driving companies to work together so as to spread and reduce the risks. Recently, in the mobile computing industry, various end-user devices were introduced like UMPC (Ultra mobile PC), PDA (Personal Digital Assistant), smartphone, MID (Mobile Internet Device), tablet (iPad), Netbook, Ultrabook and Notebook. However, none of them has been well accepted as the final solution for customers. In the following part, a deep-study on three types of uncertainty issues is conducted (Rong, Shi, et al. 2013; Rong, Lin, et al. 2013).

2.5.1   Market uncertainty

It is well accepted that in the future people will be connected by the Internet everywhere, anytime. People would like to use mobile computing devices when they are engaged in activities such as communication, business, travelling, entertainment, education, social activities and exercise. What kinds of functions future mobile devices may have is still a big question. Regarding different activities, mobile devices should have relevant functions to meet the uncertain market requirements. For example, when people are doing business, possibly they need a video chat. So the mobile device should be able to connect with the Internet, online camera and be portable for daily business trips (Rong, Shi, et al. 2013). Moreover, people also need to use mobile for the social events connecting friends. However, mobile devices do not all need all functions. Hence, companies are supposed to design appropriate devices to deal with the uncertainty.

2.5.2 Application uncertainty

To meet potential market requirements, mobile devices need to be able to be always on (operating time long enough for several days), available with quick reboot capability, connected (always has Internet connection). There are two groups of solutions in Table 2.2, however, they failed to meet with such requirements.

In terms of the first group, Intel-based devices like the notebook, ultrabook and netbook were derived from the PC industry. The netbook evolved as a cheaper and lighter notebook providing general computing and web-based applications. It was pioneered by one OEM, and became popular due to its interesting design and portability. However, the battery life at the time was too short to meet people's daily use, and there was also a lack of 3G networks. The ultrabook could be regarded as the improved version of the netbook; however it still had problems in portability, long-time batteries and low cost.

In terms of the second group, the smartphones, smartbook, the UMPCs, MIDs and iPad were mainly based on ARM's architecture. The iPad was delivered by Apple in order to occupy the entertainment market of the mobile computing industry. However, this device is still not convenient for doing simple computing. A smartbook had features like smartphones as well as full keyboards. Smartphone was developed from the 2G mobile phone but with online applications. Intel proposed the concept of UMPC and MID with its partners in order to provide a multimedia-capable handheld computer. However, neither UMPC nor MID provided an ideal solution for entertainment compared with iPad. As a result, iPad occupied more market share.

The two competing groups with Intel or ARM chips tried to penetrate each other's market area and the boundaries between the device types became blurred (Rong, Shi, et al. 2013). From the discussion above it is clear that there is as yet no perfect solution for the final device that meets the features of all-day-on, available, connected, easy operation as well as low cost. The two groups of companies already overlapped, and were busy working out the potential device.

2.5.3 Architecture/platform uncertainty

So far there are roughly two choices of chips' architecture to deal with application uncertainty: one is the Intel architecture from the PC industry, and the other is the ARM architecture from the mobile industry. In the PC industry, the basic technology architecture is proposed by Intel which has already dominated the PC industry as a processor manufacturer. In the mobile industry, ARM's architecture is the most popular and already has a 98 per cent market share as the basic architecture of mobile phone processors (www.ARM.com). Previously the two companies never competed with each other, because they were positioned in different industries. It was said that the recent trend of convergence of PC and mobile compelled them to compete face to face.

As seen from Table 2.4, ARM's architecture aims at the mobile world with the advantage of low power, small size and high integration. Nevertheless, because the final devices are different from one another, there is an implication that the key problem to be overcome is that they should be compatible. Intel has the dominant design for the PC industry where the standardisation process is well achieved. However, when the issue is one of high computing requirement, power consumption is usually high, and the size is not small enough for portable use.

As a result, the problem is 'What is the right architecture to develop a final device in this emerging industry?' The currently available architectures have both advantages and disadvantages. Each group has to consider how to solve their problems to improve their architecture performance. This problem also puts their partners in a complicated situation. For example, Wistron, as a famous ODM, already developed a MID based on both of these two kinds of architecture. But because of their advantages and disadvantages, Wistron was in a dilemma. Symbian was different as a powerful player in the mobile world, so that it still preferred to develop the OS based on the ARM's architecture. Regarding the OEM companies, they definitely would try both platforms, as the situation was not clear enough.

Table 2.4   Comparison of different architectures

	ARM	Intel	Ideal Requirements
Power consumption	Low	High	Low
Performance	Good	Excellent	Excellent
Size	Small	Big	Small
Compatible	Normally	Excellent	Excellent
Integration	High	Medium	High
Instruction set Architecture	RISC	CISC	Not yet

2.6 Key industrial challenges: interoperability

As the future is uncertain, industry requires a high degree of interoperability among partners in order to reduce risk and to approach the best design. Exploratory cases show that as the end-user products became very complicated and uncertain, partners began to collaborate with one another more than ever before, which finally resulted in the high degree of partners' interoperability as Table 2.5.

2.6.1 Company strategy: platform interoperability

The focal/core firms or keystone companies (Iansiti & Levien 2004b; Adner & Kapoor 2010) in the mobile computing ecosystem are those that have the technology platform and encourage partners to develop new products (Gawer & Cusumano 2013) based on the platforms. Normally, products in the mobile computing industry are complicated and vary both on the technology and market sides. However, it is also a big challenge for focal companies to persuade and coordinate all their supply-chain partners to adopt their platforms. As a result, focal companies have to collaborate with all parts of the mobile computing industry as much as possible in order to promote their platform and architecture.

Table 2.5 Different companies' strategies

	Case type	Core company	Non-core company
Symbian	OS		Several OEMs help build OSV
Tecent	ISV		Developed based on different OS platform
TSMC	Foundry	Design for manufacturing platform collaborating with EDA companies, architecture providers.	
China Mobile	Carrier	OS platform, and support and collaborate all the value chain	
VIA	IC design	Integrated supply chain, set up netbook platform	
Wistron	ODM		Cooperate with IC and provide total solution for OEM
Apple	OEM	Collaborate with all parts of value chain, acquire trend, investment direction	

Meanwhile, companies without their own platform form small alliances and offer products based on one or more of the various existing platforms.

As shown in Table 2.5, **TSMC**, in order to shorten manufacturing lead time, ensure manufacturing feasibility, as well as attract more customers, organised a platform called DFM – 'design for manufacturing ecosystem'. In this DFM, TSMC shared functional IPs of manufacturing feasibility with design tool companies, IP providers and IC design firms, and also encouraged them to contribute their IPs.

Then, as the mobile computing industry emerged, services and applications were regarded as the future growth areas. **China Mobile** was willing to penetrate into these areas rather than only act as a tube carrier. So they initiated a platform with an open-source operating system, encouraging partners to adopt their OS and to develop compatible applications and services.

VIA set up an integrated supply chain to provide a total solution for the future production of netbooks, with two purposes: to reduce the risk of dealing with market uncertainty, and to compete with the huge player – Intel.

By learning from the failure of vertical integration in the PC industry, **Apple** became more active in interacting with partners and customers. They built up the online store which achieved win/win/win business for Apple, their partners and their customers, stimulating contributions to an online application community.

2.6.2   Company strategy: supplementary interoperability

Rather than being focal/core companies, the non-core companies in the mobile computing ecosystem are those that do not have a technology platform but have specialised capability to support focal companies' platforms to develop new products.

In **Symbian**'s case, to take advantage of the new opportunities of smartphone OS and maintain competitive advantage beyond competitors, they were made into an open-source resource by Nokia, which encouraged different OEMs, OSVs and ISVs to work on their OSs. Nokia made their mobile phone diversified by adopting these OSs coming from the open source.

In terms of **Tecent**, they had to deal with more than 20 different OS platforms. So they built up small alliances with their customers to generalise their products and make them compatible with different operating systems.

Wistron was an ODM-like service provider which initiated alliances to develop total solutions for new product development. The OEMs and carriers would get this total solution instead of directly discussing the specification of new products with many non-core companies. So far Wistron has also expanded their direct customers into three types (Mobile phone OEM, PC OEM and Carriers) to deal with market uncertainty. All these three customers hoped to penetrate into the mobile computing end-user product market even though they came from different sectors.

These cases indicate that in order to win competitive advantage, non-core companies usually acted in supplementary roles to core companies' platforms by integrating their capabilities.

2.7   Conclusion

This chapter explored the recent and ongoing development of the mobile computing industry from the semiconductor industry. It also identified the key industrial challenges by studying seven exploratory cases. Furthermore, it highlighted two industrial challenges as 'uncertainty' and 'interoperability'. The industrial studies suggest there is a big potential opportunity to question the existing established theories and to tackle new theory around the general question: 'In the recent and current dynamic environment, how might firms rapidly capture and commercialise ideas, maintain the process sustainably, and finally successfully nurture a new industry?'

3
Literature Review

3.1 Introduction

This chapter will re-examine how the relevant manufacturing system theories tackle the emerging industrial challenges: uncertainty and interoperability which identify the research gaps – the potential and demanding area as the business ecosystem theory. Hence, the business ecosystems research will be systematically reviewed since the theory addressed well the industrial challenges.

Thus this chapter is structured in three steps:

1) to study the manufacturing system regarding the industrial challenges – uncertainty and interoperability, and to highlight potential research areas in business ecosystems;
2) to review previous studies of business ecosystems and related issues;
3) to discover what research gaps may not have been covered in previous research about business ecosystems.

3.2 Manufacturing system theories of industrial challenges

3.2.1 Review scope

Manufacturing systems have been studied at four levels: firm, intra-firm (international manufacturing network), inter-firm (supply chain) and global supply chain (internationalised inter-firm). Shi and Gregory (2003) show three trends of current manufacturing system evolution. In the first trend, companies concentrate simply on internationalisation and intra-firm plant network dispersion. The second trend focuses on ownership externalisation. Companies outsource non-competitive parts to their partners to form a supply chain. The last trend combines the previous two, having characteristics both of ownership expansion and geographic dispersion. This forms a new type of manufacturing system: a global manufacturing virtual network (GMVN) – a new operational environment is required to allow manufacturing systems to access, optimise and deploy strategic resources (Shi et al. 2003).

Based on Shi and Gregory's (1998) matrix, other relevant theories are positioned in each square. On the inter-firm level, in order to improve efficiency and rapid response, companies along the supply chain will form clusters to improve the quality of interaction (Bergman & Feser 1999a). Besides industry clusters, a business network is regarded as a set of connected business relationships in which resources are exchanged and commitments made to fulfil value creation (Astley & Fombrun 1983). At the internationalised inter-firm level, international strategic alliances extend the scope of coordination into the international inter-firm level, while open innovation embraces ideas from both the R&D and the marketing sides (Chesbrough 2003), dealing with the new operational environment in ways similar to the GMVN context.

In the following sections, the literature on level of firm, intra-firm, inter-firm and internationalised inter-firm will be reviewed to demonstrate whether those theories are explored well enough on the two industrial challenges: uncertainty and interoperability. The review also highlights potential research areas for the business ecosystem study.

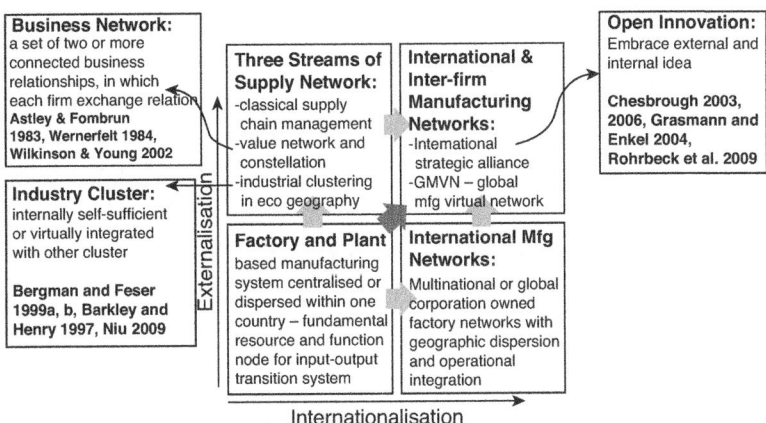

Figure 3.1   The relevant theories to business ecosystem
Source: Adapted from Shi et al. (2003).

3.2.2   Manufacturing system: firm level

System configuration has a big impact on manufacturing strategy. Hayes and Wheelwright (1984) presented a 'structural–infrastructural' framework with eight key elements to describe firm-based manufacturing systems and decision-making. Four of the key elements: capacity, facilities, technology and vertical integration are normally regarded as structural factors which tend to be of comparatively long-term impact. The other four elements – workforce, quality, production planning/materials control and organisation – are viewed as the infrastructural parts because they are 'tactical' in nature and include many areas requiring ongoing decision-making. These eight elements together form a whole system for operational decision-making (Hayes & Wheelwright 1984).

Regarding market uncertainty, Hayes and Wheelwright proposed a best-practice model by matching system configuration with its manufacturing process (Hayes & Wheelwright 1979, 1984). This model categorised different patterns of manufacturing system such as project-based, job flow, batch, line flow and machine-paced flow. Each pattern demonstrated different manufacturing processes and relevant product markets. For example, the project-based system was good at manufacturing customised product with slow response and low volume, while the machine-paced flow system did well in producing a high volume of products with fewer customised features. In conclusion, these strategies at the firm level offered a way to configure systems and to adapt to different markets. However, a firm-based manufacturing system was not flexible and agile enough to cope with the changing nature of industry.

The most influential contribution from Hayes and Wheelwright was that they researched, analysed and deconstructed manufacturing systems to provide a useful set of descriptive constructive elements for manufacturing systems and strategies. Their ideas have also been applied to manufacturing systems at intra-firm, inter-firm and internationalised inter-firm levels.

3.2.3   Manufacturing system: intra-firm level

The system constructs, manufacturing process and its configuration pattern were highlighted at the firm level (Hayes & Wheelwright 1984). As the manufacturing system evolves or grows into an intra-firm manufacturing network, those three key parts of the system also vary (Shi & Gregory 1998). Ferdows proposed a view of the international manufacturing system as a network of factories with the primary strategic rationales of access to low-cost production, skills/knowledge and proximity to market (Ferdows 1989a, 1989b). Furthermore, Shi and Gregory (1998) suggested that international manufacturing networks (IMN) were multi-nationally dispersed factory systems for a product family or a strategic business unit. Following Hayes and Wheelright's framework, Shi and Gregory also investigated IMNs and explored their manufacturing strategies. The structural factors of an IMN system were determinants settling the various patterns of the system, while infrastructural factors were dynamic controllers from daily operations and accumulative improvement to product transfer and network evolution during internationalisation (Shi & Gregory 1998).

Shi and Gregory (1998) also defined the configuration patterns used by IMNs to cope with the uncertain nature of industry. They used dimensions of geographic dispersion and manufacturing coordination to categorise IMNs into seven patterns to achieve four strategies: multinational strategy, international strategy, global strategy and transnational strategy (Ghoshal & Bartlett 1990).

These strategies, such as 'configuration for markets', were similar to those in firm-level manufacturing systems and were not flexible enough for the dynamic nature of industry.

3.2.4   Manufacturing system: inter-firm level

Supply-chain management

The term 'supply-chain management' expresses the need to integrate key business processes, from supplier to end-user by exchanging information between the production and the marketing functions (Oliver & Webber 1982). During the past decades, due to the process of globalisation and localisation, the uncertainty in supply-chain management was highlighted. It was regarded as one of the key issues that impact on the effectiveness of a supply chain and the performance of the manufacturing function (Davis 1993; Van der Vorst & Beulens 2002).

Three main streams of uncertainty were also suggested: first, inherent characteristics cause fluctuation of demand, process and supply (Blackhurst et al. 2004); second, the fundamental factors of supply-chain configuration result in potential disturbances (Harland 1996); and third, external phenomena such as markets, technology and governmental regulations disturb the supply-chain system (Wilding 1998). Also, appropriate supply-chain redesign strategies were developed in order to reduce the internal and external uncertainty. Supply-chain configuration, control structure, information system, and organisation and governance structure should be redesigned to be more adaptable to uncertainty (Van der Vorst & Beulens 2002). Van der Vorst and his colleagues also suggested combining two levels of operations: strategic configuration and operational management (Van der Vorst et al. 2006).

In terms of supply-chain uncertainty, academic researchers mostly focus on channels of information flow and material flow (Christopher 1999; Lambert & Cooper 2000; Thomas & Griffin 1996) with the purpose of achieving order fulfilment, customer-relationship management and product development. In conclusion, the study of uncertainty in supply chains does not focus on nurturing new and uncertain markets but on maintaining production for existing markets.

Business network theory

Business network theory describes a set of two or more connected business relationships (Emerson 1976). The Resource-Based View (RBV) and transaction cost (TC) theories helped to describe why business networks are formed (Chen & Chang 2004). The RBV theory argues that firms that possess resources can achieve competitive advantage (Wernerfelt 1984), while transaction costs are the main concern when a company has to choose between producing internally or acquiring (Williamson 1995).

Later on, the ideas of business networks and relationships were developed to solve issues of relationship commitment, interdependence, value creation (Holm et al. 1999) and relationship structure. Generally, as ICT technology emerged, one-to-one relationships among firms tended to be replaced by multiple and interdependent relationships. Firms were engaged directly or indirectly in several markets. In order to get access to complementary resources and capabilities, companies became more interdependent and business relationships became more essential (Williamson 1995). The purpose of studying and building relationships was to achieve value creation by coordinating different partners and resources (Cook & Emerson 1978). Relationship structure had a big influence on the innovation process. More embeddedness was suitable for incremental innovation, while less embeddedness was feasible for radical innovation (Holm et al. 1999).

In conclusion, business network theory focuses on relationship development for resource capture and value creation rather than coping with market uncertainty and nurturing.

Industry clusters

Industry clusters are geographic concentrations of interconnected companies and institutions in a particular field (Porter 1990). Roles inside clusters not only aim to ensure competitive advantage in a global economy but also draw increasing value from local knowledge, relationships and motivations (Porter 1998). Industry clusters are geographically and economically sensitive, particularly at policy level (Bergman & Feser 1999b).

Initially a cluster is more geographically oriented to improve industry efficiency (Scott 1992). Later, in order to share common knowledge and complementary capability, companies start to form new clusters that are more industry-oriented and based on specialised knowledge and resources (Porter 1990). With the help of ICT technology, companies began to form virtual networks and to add value to the value-creation network by exchanging digital knowledge with other members (Passiante & Secundo 2002).

Besides these cluster types, scholars also began to study individual firm's performance in clusters, the relationships among firms, and how the structure of clusters relates to their innovation activities (Bell 2005). However, firms in clusters mostly had similar aims of achieving industry efficiency and reducing innovation risk, but paid less attention to product variety. The interactions were mostly focused on forming a common value chain.

In conclusion, the networking strategy of industry clusters took efficiency improvement and collective activity as their first priority rather than innovation to meet an uncertain future.

3.2.5 Manufacturing system: internationalised inter-firm level

International Strategic Alliance (ISA)

Child and Faulkner (1998) presented a definition about strategic alliances: 'cooperative strategy is the attempt by organisations to realise their objectives through cooperation with other organisations, rather than through competition with them'. ISA, a typical form of business relationship, is developed in order to achieve global economies of scale and leanness, to share complementary resources and to reduce risk and help shape the international market (Porter & Fuller 1986; Hamel 1991). There are three dimensions generally used to categorise strategic alliances: scope of alliance; joint venture or non-joint; and two partners or a consortium (Faulkner 1995). The scope of alliance can range from focused to complex. Joint or non-joint venture refers to whether it is collaboration with or without a strategic purpose. A consortium alliance is where there are more than two alliance partners. A strategic alliance normally has a set agreement goal which also defines the orientation of a stable relationship.

ISA theories were developed to describe the processes of globalisation and localisation (Koza & Lewin 1998). After the 1980s, global competition highlighted asymmetries in the skill endowments of firms (Hamel 1991), and strategic alliances were implemented across borders. In the 2000s, Shi investigated whether ISAs, including intra-firm coordination and inter-firm cooperation, included a wide span of long-term commitment towards virtual community or traditional relationships for a flexible environment (Li et al. 2000). Four criteria were also identified to help categorise ISAs: prior international alliance experience; administrative governance form; nationality of foreign partner; and motive for alliance formation (Nielsen 2003).

Regarding the future direction of application or industry trend, ISA mostly focuses on knowledge transfer which is developed from a resource-based view (Mowery et al. 1996). Hermens proposed a framework about leveraging knowledge, discovering complementarities between the technologies and activities of partners to meet the uncertainty of industry trends (Hermens 2003). Furthermore, the effects of knowledge ambiguity were also studied in terms of knowledge transfer to understand the firm's level of collaborative know-how, its learning capacity and the duration of the alliance (Simonin 1999).

In conclusion, ISA mostly focuses on building stable relationships based around complementary resources and partners with a set agreement goal.

Global manufacturing virtual network (GMVN)

GMVN is defined as a network of inter-firm partnerships – typically limited life and evolutionary dynamics – between globally distributed and collaborating companies which agree to collaboratively exploit a particular market opportunity via their core competencies and resources (Li et al. 2000). GMVN was developed from global manufacturing networks (GMN) based on resource ownership. It extended firms' boundaries in order to coordinate and leverage resources like a virtual community (Shi et al. 2003).

GMVN indicated strategic collaboration, including strategic alliances for long-term commitment towards a virtual community or arm's-length trading relationship development for more flexibility (Shi et al. 2003). There were two ways to highlight GMVN's successful collaborative manufacturing in comparison with virtual organisations and strategic alliances: one was collaboration between two or more partners in order to deliver the products and services. The other was about the business environment, where firms with unique core competence were eager to collaborate with others to form a supply chain. The more open the environment was, the better would be the collaboration triggered (Shi & Gregory 2005).

With these two advantages, Shi also proposed a process life-cycle model of GMVNs Figure 3.2. GMVN emerged from inspired innovators, system integrators and customer solution providers, who exploited ideas towards a dynamic and diversified business reality. In its early stages, the mission of a GMVN was to enable innovation to be quickly and cheaply transferred into a successful commercial product. In its later stages, it might evolve into a more efficient enterprise or dissolve to allow the component companies to re-enter the original GMVN environment for future collaboration (Shi 2004; Shi & Gregory 2005).

GMVN studies concentrate on the coordination of resources and the delivery of products or services, but lack information on the sustainability of companies' futures, the nurturing of business environments and coping with the uncertainty of industry trends. How to nurture an industry and to sustain innovation is still a big challenge for GMVN research.

Open innovation

The concept of open innovation was first introduced by Chesbrough (2003) as a paradigm assuming that firms can and should use external ideas as well as internal ideas and internal and external paths to market. Open innovation emerged from closed innovation systems based on six driving forces: (1) globalisation; (2) technology upgrading; (3) shortening of product life cycle; (4) risk and integration of innovation; (5) complexity of innovation; and (6) R&D resource (De Backer 2008).

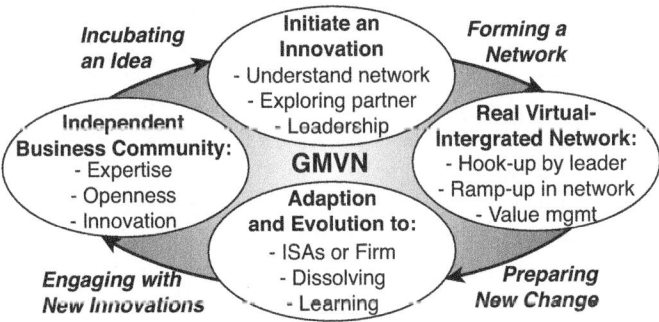

Figure 3.2   GMVN life cycle

Academic researchers increasingly tried to understand the innovation process in order to inform technical and market development. Chesbrough first suggested four ways to meet the demand: information-based capital investment; early exposure of failure; out-licensing rejected projects; and spin-off technology (Chesbrough 2004). The open innovation process has also been divided into three archetypes: the outside-in process; inside-out process; and coupled process from inbound and outbound perspectives (Gassmann & Enkel 2004). Furthermore, a more general innovation process was identified as idea generation, research, development and commercialisation. In each part there were also relevant instruments to implement each process, including foresight workshops, executive forums, customer integration, endowed chairs, consortia projects, corporate venture capital, Internet platform, joint development, strategic alliance, spin-out and market testing (Rohrbeck et al. 2009).

In Cambridge University, Minshall and his colleagues studied open innovation in detail by exploring the partnership model between start-up and large firms when large firms implemented open innovation strategy (Minshall et al. 2008). They tested the asymmetric partnership and provided strategies over five dimensions: business strategy model; technology; organisation; deal set; and deal-ongoing management. (Minshall et al. 2010). They also studied open innovation from the organisational culture perspective to understand the R&D subculture, and to provide guidance for blue-sky and applied R&D (Mortara et al. 2010).

Open innovation would encourage three types of activities: user innovation, market leading and open source (De Backer 2008). User innovation was more demanding and involved companies' new product development. Companies that led the market would encourage latecomers to develop their own product based on the focal company's platform. More and more companies began to make value from open source information which is nurtured by non-profit individuals or groups. All of these activities show ways in which open innovation network partners cope with new market demand and uncertainty. However, open innovation is focused mostly at a strategic level on R&D and market requirements, not on the whole value chain, which is limited when compared with the global manufacturing virtual network (Li et al. 2000). For example, questions of how to coordinate a manufacturing network, how to persuade other partners to be involved, how to configure their roadmap, all need more exploration and research.

3.2.6 Conclusion of literature review of uncertainty and interoperability

The eight theories described have been scored as to how well they relate to the two industrial challenges (Uncertainty and Interoperability) and shown in Table 3.1. In the sub-dimensions of uncertainty, the scoring is: stability (1), some uncertainty (2) and uncertainty (3). In the sub-dimensions of interoperability, the scoring is integrated (1), less interaction (2), high interaction (3).

In the table, the score for each sub-dimension is summed to give an indication of how well each theory addresses the challenges. To visualise the result, the theories are shown in a two-dimensional matrix of uncertainty and interoperability (Figure 3.3).

The firm-level and intra-firm–level systems both used the configuration methods to divide existing market into different types. Both theories were unable to cope with requirements from an uncertain market and dynamic innovation. Instead they dealt with a classification of fixed markets.

At the inter-firm level, the supply chain is composed of partners who cooperate to deal with the uncertainties of information and material flow. Business network theory studies a set of relationships and does include elements of market uncertainty and partner interoperability. Industry cluster theory studies the geographic concentration of interconnected companies and institutions in a particular field. At a later

Table 3.1 Theories on industrial challenges

Challenges Theories	Uncertainty				Interoperability		
	Market	Application	Technology	Mark	Platform	Supplementary	Mark
Firm level	1	1	1	3	1	1	2
IMN	2	1	1	4	1	2	3
Supply chain	2	2	1	5	1	3	4
Business network	2	2	1	5	2	3	5
Industry cluster	2	1	1	4	2	3	5
Strategic alliance	2	2	2	6	3	2	5
GMVN	3	3	1	7	3	3	6
Open innovation	3	3	3	9	2	2	4

stage, they also proposed a cluster of complementary resources with a knowledge transfer orientation and thus touch on interoperability and application and market uncertainty.

At the internationalised inter-firm level, an international strategic alliance is the long-term agreement between two or more partners for a specific market in order to get complementary resources and to reduce risk. This study also touches on platform-based interoperability. GMVN is a virtual networked manufacturing system, which is not oriented by location, but addresses the dynamic environment and coordinates worldwide resources. A GMVN initiates an innovation and coordinates complementary resources and thus already covers a big part of uncertainty and interoperability. Open innovation, as it addresses external and internal idea exchange in both R&D and marketing, so focuses greatly on the uncertainty part.

However, after positioning these theories and disciplines in the matrix, none of them completely address the study of both uncertainty and interoperability, hence the top-right point of Figure 3.3 requires new thinking to meet the new demands of uncertainty and interoperability research. In Chapter 2, industrial practitioners have already highlighted

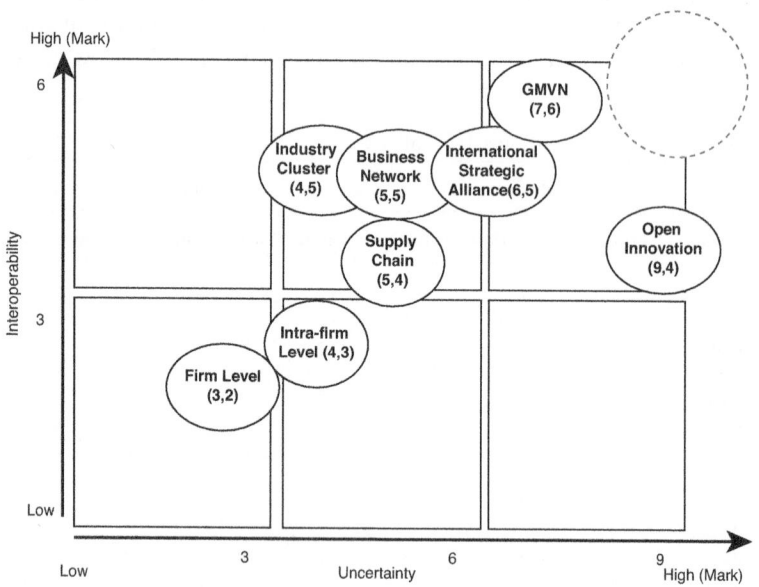

Figure 3.3   Conclusion of literature review

the importance of the business ecosystem and have been making efforts to build up the business ecosystem. Figure 3.3 highlights the potential power of the concept 'business ecosystem'. The business ecosystem concept was first proposed by Moore in 1993 as an economic community where different roles co-evolve with one another in a business environment. Moore paid much attention to the areas of uncertainty (business environment) and interoperability (co-evolution) (Moore 1993). In the following section, an in-depth, systematic exploration of the business ecosystem idea is carried out in order to identify research gaps.

3.3 Review of business ecosystem theories

3.3.1 The definition of business ecosystem

Table 3.2 lists the most representative studies of academic papers, reports and books exactly addressing the definition of the business ecosystem area since 1993. The definitions, from different researchers, vary (Anggraeni et al. 2007).

Moore first proposed the business ecosystem in 1993 and 1996 (Moore 1993, 1996):

> An economic community supported by a foundation of interacting organisations and individuals – the organisms of the business world. This economic community produces goods and services of value to customers, who are themselves members of the ecosystem. The member organisations also include suppliers, lead producers, competitors, and other stakeholders. Over time, they co-evolve their capabilities and roles, and tend to align themselves with the directions set by one or more central companies. Those companies holding leadership roles may change over time, but the function of ecosystem leader is valued by the community because it enables members to move toward shared visions to align their investments and to find mutually supportive roles.

In Moore's view, his work aimed to describe how the economic community worked by highlighting the interaction between companies and their business environment which included extended enterprises (non-direct business partners around core business), other levels of organisations, competitors and business opportunities as seen in Figure 3.4.

Following Moore's definition, Iansiti and Levien (2002) regarded the business ecosystem as a large number of loosely interconnected participants who depended on one another and had a shared fate. Furthermore,

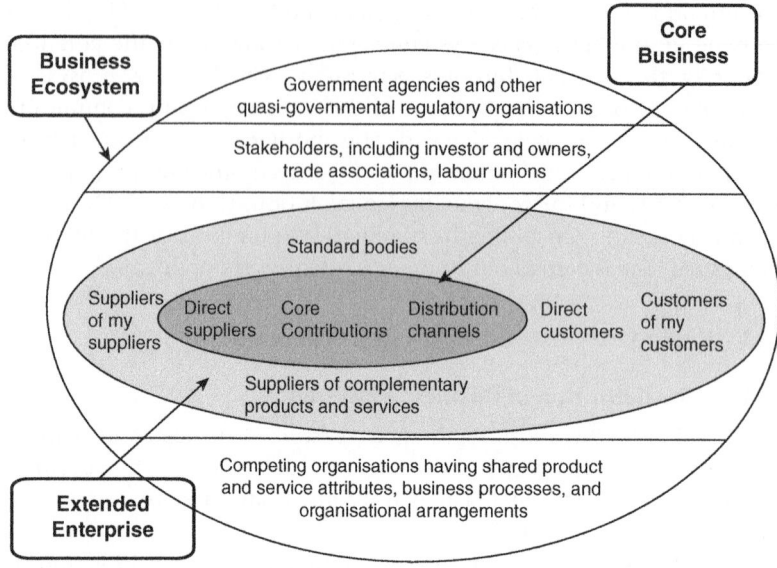

Figure 3.4   The whole picture of a business ecosystem
Source: Moore (1996).

they developed different roles to implement their strategies inside this network. Then a systematic picture of how a business ecosystem operated was developed to emphasise the interconnection of competition and cooperation, and highlight their shared-fate process (Peltoniemi & Vuori 2004; Peltoniemi 2005, 2006). From the literature, some key points can be identified: the space of opportunities – business environment (Kraemer & Dedrick 2002; Fragidis et al. 2007; Moore 1993, 1996; Gueguen et al. 2006); community of interdependent organisations (Chang & Uden 2008; Desai et al. 2007; Den Hartigh & Van Asseldonk 2004; Den Hartigh et al. 2006); co-evolution with visions (Marín et al. 2007; Peltoniemi & Vuori 2004; Desai et al. 2007).

Other works later continued to highlight the importance of these three key ideas as well as to add more novel perspectives. With the ICT technology boosting, technology factors had more impact on the business ecosystem operation (Power & Jerjian 2001; Adomavicius et al. 2006; Iansiti & Richards 2006). However, they mostly discussed technological tools instead of the nature of a business ecosystem. Some researchers also highlighted customer views of the business ecosystem and renamed it as customer ecosystem (Fragidis et al. 2007; Manning et al. 2002).

Table 3.2 Overview of the business ecosystem definitions

No.	Author & Year	How they describe the business ecosystem
1.	Moore 1993	Like its biological counterpart, gradually moves from a random collection of elements to a more structured community.
2.	Moore 1996	An economic community supported by a foundation of interacting organisations and individuals – the organisms of the business world. This economic community produces goods and services of value to customers, who are themselves members of the ecosystem. The member organisations also include suppliers, lead producers, competitors, and other stakeholders. Over time, they co-evolve their capabilities and roles, and tend to align themselves with the directions set by one or more central companies. Those companies holding leadership roles may change over time, but the function of ecosystem leader is valued by the community because it enables members to move toward shared visions to align their investments and to find mutually supportive roles.
3.	Power & Jerjian 2001	A system of websites occupying the world wide web, together with those aspects of the real world with which they interact. It is a physical community considered together with the non-living factors of its environment as a unit.
4.	Farhoomand et al. 2001	Where business transactions across corporate, industry, and national boundaries are conducted with an increasing degree of speed, openness and transparency.
5.	Iansiti & Levien 2002	Characterised by a large number of loosely interconnected participants who depend on each other for their mutual effectiveness and survival.
6.	Kraemer & Dedrick 2002	Define business ecology as a broad community of firms and individuals that add value to a technology standard by supplying complementary assets to the core products.
7.	Manning et al. 2002	Customer ecosystem: A complex grouping of companies and customers, suppliers, and partners that gain mutual benefit from one another.
8.	Iansiti & Levien 2004b	A business ecosystem is a business network, which is formed by large, loosely connected networks of entities, that interact with each other in complex ways, and the health and performance of a firm is dependent on the health and performance of the whole.
9.	Iansiti & Levien 2004b	Characterised by a large number of loosely interconnected participants who depend on each other for their mutual effectiveness and survival.

Continued

Table 3.2   Continued

No.	Author & Year	How they describe the business ecosystem
10.	Petoniemi & Vuori 2004	A dynamic structure which consists of an interconnected population of organisations. These organisations can be small firms, large corporations, universities, research centres, public sector organisations, and other parties which influence the system.
11.	Den Hartigh & Van Asseldonk 2004	A network of suppliers and customers around a core technology, who depend on each other for their success and survival.
12.	Quaadgras 2005	A set of complex products and services made by multiple firms in which no firm is dominant.
13.	Peltoniemi 2005	Interconnected; both competition and cooperation; can also be defined in terms of landscapes. Consists of a large number of participants that can be firms and other organisations.
14.	Peltoniemi 2006	A business ecosystem consists of a large number of participants, which can be business firms and other organisations. They are interconnected in the sense that they have an effect on each other. Interconnectedness enables various interactions between the members. These interactions can be both competitive and cooperative. Together with interconnectedness they lead a shared fate among the organisations. The members are dependent on each other, and the failures of firms can result in failures of other firms.
15.	Iansiti & Richards 2006	IT ecosystem is characterised by a large number of participants who depend on each other for their mutual effectiveness and survival.
16.	Moore 2006	A new organisation form besides market and hierarchy. Be conceived as a network of interdependent niches that in turn are occupied by organisations.
17.	Adner 2006	The collaborative arrangements through which firms combine their individual offerings into a coherent, customer-facing solution. Innovation ecosystem has become a core element in the growth strategies of firms in a wide range of industries.
18.	Iyer et al. 2006	Virtual integration through alliances to establish networks of influence and interoperability.
19.	Gueguen et al. 2006	Business ecosystems are based on the idea that several companies will collaborate to improve their offer: importance of leadership, the role of 'keystone organisations', the principle of co-evolution, the 'coopetition' dynamics and evolution in the rules of competitiveness, the ability to develop in accordance with an open-source community, the diversity of the key players.

Continued

Table 3.2   Continued

No.	Author & Year	How they describe the business ecosystem
20.	Adomavicius et al. 2006	The term technology ecosystem is appropriate because it emphasises the inherently organic nature of technology development and innovation that is often absent in standard forecasting and analytical methods.
21.	Fragidis et al. 2007	Business ecosystems concentrate large populations of different kinds of business entities. They transcend industry and supply chain boundaries and assemble a variety of organisations that can complement each other and synergistically produce composite products. Interdependence and symbiotic relationships are inherent attributes in business ecosystems; as a result, the participants counter a mutual fate and co-evolve with each other. But in parallel, members compete with each other for the acquirement of resources and the attraction of customers.
22.	Marín et al. 2007	To flourish in such environments, businesses must continually adapt and evolve. This requires that a business engage in an ongoing dialogue with its environment and with others with which it shares this environment.
23.	Desai et al. 2007	Dynamic, customizable groups of services provided and used by membership-based social or business networks of varying scale and lifetime.
24.	Anggraeni et.al 2007	The business ecosystem perspective offers a new way to obtain a holistic view of the business network and the relationships and mechanisms that are shaping it, while including the roles and strategies of the individual actors that are a part of these networks.
25.	Chang & Uden 2008	A business ecosystem is a network of buyers, suppliers and makers of related products or services and their socio-economic environment that includes institutional and regulatory framework.
26.	Li 2009	A business ecosystem provides a new perspective for repositioning a company's strategy in order to aggressively further its own interests and to promote its overall ecosystem health.
27.	Adner & Kapoor 2010	An innovation ecosystem comprises not only the core innovator, but also its upstream suppliers, and its downstream buyers and complementors.
28.	Ceccagnoli et al. 2012	Platform ecosystem: the network of innovation to produce complements that make a platform more valuable
29.	Williamson and De Meyer 2012	A network of organisations and individuals that co-evolve their capabilities and roles and align their investments so as to create additional value and/or improve efficiency

Continued

Table 3.2   Continued

No.	Author & Year	How they describe the business ecosystem
30.	Gawer and Cusumano 2013	External (industry) platforms as products, services, or technologies that are similar in some ways to the former but provide the foundation upon which outside firms (organised as a 'business ecosystem') can develop their own complementary products, technologies, or services.
	Concept Conclusion	A business ecosystem is a community consisting of different levels of interdependent organisations which generates co-evolution between partners and their business environment.

From the review of business ecosystem definitions, three key phrases were identified: community of interdependent organisations; business environment (opportunity space); platform and co-evolution.

The term 'community of interdependent organisations' means the relationship among network partners is no longer simply that of supplier–customer, but that such organisations are dependent on one another and share in a common fate (Moore 1996; Power & Jerjian 2001; Iansiti & Levien 2002; Kraemer & Dedrick 2002; Peltoniemi & Vuori 2004; Den Hartigh & Van Asseldonk 2004; Iansiti & Levien 2004b; Moore 2006; Iyer et al. 2006; Iansiti & Richards 2006; Fragidis et al. 2007; Desai et al. 2007; Chang & Uden 2008).

The term 'business environment' means that organisations in a business ecosystem should expand their views beyond the supply-chain partners of their core business. The business environment includes other non-direct business partners such as government agencies, industry associations, stakeholders and also competitors who shape the industry greatly (Moore 1993, 1996; Gueguen et al. 2006; Marín et al. 2007). The business environment can also be regarded as an opportunity space in which the interdependent organisations share their ideas and visions for future development.

The term 'co-evolution' refers to the idea that interdependent organisations will co-evolve with one another and with the dynamic business environment, and highlights the importance of key firms' interactions with their business environment as well as with core business partners. Normally there will be an ecosystem platform to facilitate the focal firm and complementors to achieve collective behaviour and co-evolution (Adner & Kapoor 2010; Ceccagnoli et al. 2012; Gawer & Cusumano

2013). Thus organisations can share visions in order to shape the future together (Moore 1996; Farhoomand et al. 2001; Manning et al. 2002; Quaadgras 2005; Peltoniemi 2005, 2006; Adner 2006).

In order to make the study clearer and more focused, a preliminary working definition based on previous literature is delivered below:

> A business ecosystem is a community consisting of different levels of interdependent organisations that stimulate co-evolution between partners and their business environment.

3.3.2 Business ecosystem review over time

In the following section, the business ecosystem literature from 1993 onward will be reviewed, listed as Table 3.3. There are five main groups contributing to business ecosystem theory. The first is Moore who was the first to propose the concepts of business ecosystem, business ecosystem life cycle and public 'goods' (benefits) in his one book and two papers (Moore 1993, 1996, 2006). The second group is Iansiti and his colleagues who developed the business ecosystem idea based on Moore's concept and enriched it by adding different role types, their strategies and ecosystem health. So far they have published four papers mentioning those issues (Iansiti & Levien 2002, 2004a, 2004b; Iansiti & Richards 2006). The third group is Adner and his colleagues who regarded the innovation ecosystem as the structure of technology interdependence. This structure, composing of focal firm, complementors, supplier and customers, demonstrates that the focal firm's commercialisation could be improved by the support of complementors (Adner 2006, 2012; Adner & Kapoor 2010; Kapoor & Lee 2013).The fourth group is Peltoniemi and his colleagues who proposed the four key features of a business ecosystem and proposed the key governance framework by adopting system complexity and evolutionary theory. They have published three representative papers on the business ecosystem (Peltoniemi & Vuori 2004; Peltoniemi 2004; Peltoniemi 2006). The fifth group is Den Hartigh and his colleagues who worked on the new types of roles, governance framework and ecosystem health measurement at the firm level. They have published three papers (Den Hartigh & Van Asseldonk 2004; Den Hartigh et al. 2006; Anggraeni et al. 2007). Besides the areas explored by the five main groups described above, other related areas have also been researched, such as platform- and ICT-based ecosystem, standardisation and governance framework (Power & Jerjian 2001; Quaadgras 2005; Gueguen et al. 2006; Adner 2006; Fragidis et al. 2007; Iyer et al. 2006; Bannerman & Zhu 2008; Chang & Uden 2008; Ceccagnoli et al. 2012; Gawer & Cusumano 2013).

Table 3.3   Overview of the business ecosystem (BE) study since 1993

No.	Author & Year	Year	Research highlight	Detail
1.	Moore 1993, 1996	1993 1996	Driving force; concept, life cycle	Different levels of organisation co-evolution with environment; Life cycle: birth, expansion, authority and renewal
2.	Power & Jerjian 2001	2001	E-business strategy	12 principles to reconfiguration e-business
3.	Iansiti & Levien 2002, 2004a, 2004b	2002 2004	Roles and firm strategy, health measurement; foundation as strategy	Roles and their strategies (keystone, dominator, hub landlord, niche player); health measurement: productivity, robustness, niche creation; foundation: architecture, integration, market
4.	Manning et al. 2002	2002	Six steps to nurture demand ecosystem	1. Understand customer and mapping ecosystem; 2. leverage partners; 3. build adaptive strategies; 4. use information technology; 5. build ecosystem capabilities; 6. synchronise supply and demand
5.	Peltoniemi & Vuori 2004	2004	Literature review; governance framework on complexity study	Review the ecosystem from different perspectives including industrial ecosystem, digital ecosystem, et.al; features: complexity, self-organisation, emergence, co-evolution
6.	Peltoniemi 2004	2004	Literature review, innovation strategy	Distinguish the relationship between cluster, value network and business ecosystem and their relevant innovation strategy of imitating, conservative and absorptive
7.	Den Hartigh & Van Asseldonk 2004	2004	BE boundary; roles, and analysis framework	Roles (shaper, adapter, firm of reserving the right to play); governance of relation; market dynamics
8.	Quaadgras 2005	2005	Platform for alliance	Platform to common use; absorptive capacity and capabilities

Continued

Table 3.3 Continued

No.	Author & Year	Year	Research highlight	Detail
9.	Peltoniemi 2006	2006	Governance framework on complexity study and evolutionary economics study	Self-organisation, emergence, co-evolution; individual organisation; limited knowledge; selection; variation; development; external environment
10.	Iansiti & Richards 2006	2006	Health measurement framework	The framework of 'Ecosystem structure, health measurement and performance'
11.	Den Hartigh et al. 2006	2006	Health measurement	Transfer health measurement from meso-level into firm level for industrial manager use
12.	Moore 2006	2006	Eight public goods created by the business ecosystem	Customer feedback; financed; infrastructure (tech, service, design, solution); a collaboration to create a system of complementary capabilities and companies; a space for specific ecosystem; space for business opportunity; innovation trajectory; campaign (shape future)
13.	Adner 2006	2006	Risk management in BE	Three fundamental risks: initiative risks, interdependence risks and integration risks
14.	Iyer et al. 2006	2006	Role types; BE of small world network structure	Central roles(hub, broker, bridge); average degree of partners; degree of link; cluster coefficient; path length; network density
15.	Gueguen et al. 2006	2006	Collective strategy and ecosystem competences	Collective combination of different shareable resources and competencies from different key players
16.	Adomavicius et al. 2006	2006	Technology ecosystem	Concept: a focal technology within the given context. Roles of technology: component; products; support and infrastructure; Context: social and government forces, technical forces and economic forces.

Continued

Table 3.3   Continued

No.	Author & Year	Year	Research highlight	Detail
17.	Marín et al. 2007	2007	An integrated governance model	Number of agent; amount of service offer; environment size
18.	Anggraeni, Den Hartigh & Zegveld 2007	2007	Literature review	Firm; interdependent agents; community governance; network dynamic
19.	Chang & Uden 2008	2008	Governance of e-learning ecosystem	Framework: organisation structure; communication and relational mechanism; process activities
20.	Bannerman & Zhu 2008	2008	Standardisation	Standardisation as an enabler
21.	Moore 1996	2009	Ecosystem characteristics	Platform; co-evolution; symbiosis
22.	Adner & Kapoor 2010	2010	Ecosystem structure	Structure of technology independence, focal firm, component suppliers, complementors
23.	Williamson & De Meyer 2012	2012	Ecosystem advantages	Six key ways to unlock the ecosystem advantages
24.	Ceccagnoli et al. 2012	2012	Software platform ecosystem	Software vendor performance via participation in the platform ecosystem
25.	Gawer and Cusumano 2013	2013	Ecosystem platform	Internal platform and external platform

Moore in 1993, 1996, 2006

Moore (1993, 1996) identified the business ecosystem players shown in Figure 3.4 and highlighted the co-evolution among different levels of organisations, including partners in core business, extended enterprises, competitors, industry associations and government agencies, which is an extension beyond Porter's five driving forces model (Porter 1979).

Porter's (1979) model of industry organisation proposed that there are five forces which help firms identify their position in an industry with the most attractive performance and then penetrate into the industry. The five forces are: the bargaining power of suppliers; the bargaining power of customers; the threat of new entrants; the threat of substitute products or services; and current competitors (Porter 1979). These five forces only covered a part of Moore's business ecosystem: the bargaining powers of suppliers and customers were mapped as the core business value chain of Moore's model; the competitors and substitutes were also mentioned in the extended enterprise in Moore's model (Moore 1993). Porter's model was also challenged (Mintzberg et al. 1998): firms could not easily transfer from one industry to another industry even when the new industry was very attractive; the industry became more transparent and dynamic and without much bargaining power a single position or strategy would not work. Based on those arguments, Moore suggested that the industry boundary had vanished and the concept of 'an industry' should be replaced by that of the business ecosystem. The key firms inside the business ecosystem should organise their resources well and co-evolve with their network partners across industries in order to deliver more value together (Moore 1996).

Moore further pictured the business ecosystem life cycle (BELC) with four sequential phases: phase 1 – birth; phase 2 – expansion; phase 3 – authority; and phase 4 – renewal (Moore 1993) as shown in Table 3.4.

Table 3.4 The evolutionary stages of a business ecosystem

	Cooperative challenges	Competitive challenges
Phase 1: Birth	Work with customers and suppliers to define the new value proposition around a seed innovation.	Protect your ideas from others who might be working toward defining similar offers. Tie up critical lead customers, key suppliers, and important channels.
Phase 2: Expansion	Bring the new offer to a large market by working with suppliers and partners to scale up supply and to achieve maximum market coverage.	Defeat alternative implementations of similar ideas. Ensure that your approach is the market standard in its class through dominating key market segments.
Phase 3: Authorities	Provide a compelling vision for the future that encourages suppliers and customers to work together to continue improving the complete offer.	Maintain strong bargaining power in relation to other players in the ecosystem, including key customers and valued suppliers.
Phase 4: Renewal	Work with innovators to bring new ideas to the existing ecosystem.	Maintain high barriers to entry to prevent innovators from building alternative ecosystems. Maintain high customer switching costs in order to buy time to incorporate new ideas into your own products and services.

Source: Moore (1993).

In phase 1, firms watched carefully for new opportunities with the aim of setting up value chains to create value for customers; in phase 2, the business ideas would capture value for a larger number of customers. Thus this process would be able to scale up to a broader market. In phase 3, the value adding components and processes are stable and leaders set direction to encourage partners to work together. In phase 4, a new business ecosystem has emerged from the mature business communities by raising new ideas and innovations.

Moore also suggested that the concept of 'an industry' should be replaced by the concept of 'a business ecosystem', because that could represent the comprehensive situation of an across-industries collaboration (Moore 1996). It is necessary to review the previous studies on life cycle (product life cycle – PLC, technology life cycle – TLC and industry life cycle – ILC) and link them with the BELC. Table 3.5 shows the evolution of life cycle studies from the levels of product technology to industry and business ecosystems.

The term 'PLC' represents the unit sales of a product from the moment it was listed until removed from the market (Polli & Cook 1969) with four sequential stages: introduction, growth, maturity and decline (Levitt 1981). PLC studies addressed product sales and their related areas such as product innovation and process innovation, the relationship between PLC stages and its product features. However, PLC studies lack understanding of the external environment. Since the external environment has a big impact on product sales, some potential research areas are: internal factors of firms; external factors of firms within an industry; and macro-environment (Rink & Swan 1979).

The idea of 'TLC' has similarities to PLC but refers to a technology rather than an individual product. This life cycle model has six stages: technology development; technology application; application launch; application growth; technology maturity; and degraded technology (Harvey 1984). Many scholars suggested that a good way to extend the life cycle of a technology would be to transfer it internationally from developed to less developed countries (Vernon 1966; Harvey 1984). In the future more studies should be made of the two fields: patent use and development along the technology life cycle; and methods of extending the technology life cycle (Harvey 1984; Kim 2003; Haupt et al. 2007).

'ILC': industry lifecycle. Peltoniemi delivered a systematic review of ILC with its key phases, key themes and directions for future research, based on the conclusions from 216 papers on ILC and related areas (Peltoniemi 2011). He summarised key phases as 'emerging patterns' and 'transformation to industry maturity'. In an emerging pattern firms

Table 3.5 Evolution of life-cycle study: product, industry and business ecosystem

Types of life cycle	Perspective	Phases	Key themes	Future research	Key articles
Product life cycle: PLC	Product of marketing, engineering	1. Introduction, 2. Growth, 3. Maturity, 4. Decline	1. PLC curves 2. Product and process innovation 3. The scope of industry and products 4. Product concept in PLC research 5. PLC stages	1. Internal of firm 2. External of firms within industry 3. Macro-environmental	Polli & Cook 1969; Hayes & Wheelwright 1979; Rink & Swan 1979; Levitt 1981; Klepper 1996
Technology life cycle: TLC	Technology development and use	1. Technology development, 2. Technology application, 3. Application launch, 4. Application growth, 5. Technology maturity, 6. Degraded technology	1. Technology used along product life cycle, 2. Extended TLC by internationalisation	1. Patent along technology life cycle, 2. Extended life cycle of technology, 3. Technology continuation of double s-curve 4. Technology transfer	Vernon 1966; Harvey 1984; Kim 2003; Haupt et al. 2007
Industry life cycle: ILC	Firm's network within single industry	1. Emerging (technological innovation, growth of firm number, population-level learning and building of legitimacy), 2. Transition to industry maturity (dominant design, shake-out, inter-industry effects)	1. Entry and exit rates, 2. Change in the nature of innovation, 3. Survival (entry timing, pre-entry experience, innovativeness)	1. Entry and exit change along ILC phases, 2. Sufficient performance point from product to process innovation, 3. The comparison between early mover and latecomer, 4. Pre-entry experience, 5. The fuel relationship between success and innovation	216 industry life cycle relating papers concluded by (Peltoniemi 2011)
Business ecosystem life cycle: BELC	Firms of cross-industries and business environment	1. Birth, 2. Expansion, 3. Authorites, 4. Renewal	1. Co-evolution between firm and business environment 2. Illninter-industries, or regardless of industry boundary	1. Detail process of transformation along lifecycle phases	Moore 1993, 1996)

begin to introduce new technological opportunities which stimulates the entry of many other firms (Williamson 1975). Firms also define the industry's legitimate environment and begin a process of collective learning rather than competing strategically. When the industry matures (Audretsch & Feldman 1996), the dominant design emerges (Agarwal et al. 2002) and product variety decreases dramatically (Klepper 1997). These stages describe how market structure and innovation evolves in new industries with rich opportunities for both product and process innovation (Filson 2001; Klepper & Miller 1995).

Key research themes of ILC were also listed in this chapter: first, entry and exit rates varied along ILC stages; second, firms would examine the nature and relationship of product and process innovation; and finally, survival issues highlighted questions of when to enter industry, the importance of pre-entry experience (firm owning the relevant experience before entering the new industry) and firms that feature innovativeness having a greater likelihood of survival (Peltoniemi 2011).

For future research there are still five challenging questions for ILC study: 1) entry and exit changes along ILC phases; 2) the turning point from product to process innovation; 3) the comparison between early mover and latecomer; 4) pre-entry experience; 5) the argument as to whether success delivers innovation or innovation delivers success.

Peltoniemi also argued that ILC studies did not pay much attention to the interconnectedness of industries and how these would affect their own ILCs. He thought that the effect of inter-industry influence is that frequently a mature industry leads to the birth of another new industry (Peltoniemi 2011).

In terms of BELC, Moore argued that the industry boundary vanishes as firms co-evolve with their business environment (Moore 1996), which matches or even extends Peltoniemi's idea of inter-industry interaction. As a result, BELC extends the scope of industry and product life cycle ideas with the following three aspects: 1) it highlights the co-evolution between firms and their business environment; 2) it emphasises the study of inter-industry relationships; 3) it brings up the idea of the sustainability of innovation and the renewal of existing industries.

However, in comparing these three life-cycle studies, the BELC study was still at an early stage and focused more on concept building and was not as deep as the other two life-cycle studies. In the future, the transformation between different life-cycle stages should be explored further, together with related activities such as innovation, firms' interaction, penetrating strategies, entry and exit timing (Rink & Swan 1979; Peltoniemi 2011).

After the BELC study, Moore also proposed seven business issues of a developing business ecosystem such as customer (customers' interest and participation), market (market boundaries and agencies), offer (product and service architecture), business process (business process architecture), organisation (organisational architecture), stakeholder (owners and other stakeholders), and societal value and government policy. Moore argued that these seven business issues would vary along the BELC and help renew the business ecosystem itself (Moore 1996).

In 2006, Moore indicated eight public 'goods' (four tangible and four intangible benefits) delivered by the business ecosystem itself. This study also explained how firms could achieve better growth with the help of these eight public benefits. The four tangible benefits were: 1) partner network of firms and their partners with complementary capabilities; 2) a space for business opportunities; 3) contributors; 4) firm's innovation pattern and trajectory. The four intangible benefits were: campaigns for shaping the future of the ecosystem; supporting infrastructure; customer feedback; and financing (Moore 2006).

In conclusion, Moore proposed the concept of the business ecosystem, seven business issues in developing a business ecosystem as well as its life cycle and outputs. However, he has not paid as much attention to the business ecosystem itself: what elements construct a business ecosystem and how these elements help shape different kinds of business ecosystems.

Iansiti and his colleagues in 2002, 2004, 2006

While Moore thought that a business ecosystem consisted of different levels of organisations and business environment (Moore 1996), Iansiti and Levien specifically divided those organisations into four types and also discussed their functions and strategies (Iansiti & Levien 2002, 2004b). They are the keystone player, niche player, dominator and hub landlord. The keystone players set up a platform (Iansiti & Levien 2004b; Quaadgras 2005) in order to involve contributions from other players. Niche players develop specialised capabilities to add value to a business ecosystem. The dominator integrates vertically or horizontally to manage and control a large part of its network and seizes a greater part of the value. The hub landlord extracts as much value as possible from its network without directly controlling it. They thought the keystone players and niche players contributed to ecosystem health and sustainability (Iansiti & Levien 2004b). Iansiti and Levien's role-type study matched the core ideas of Ferdows's work on categorising the strategic roles of foreign factories within an international manufacturing network. This work of distinguishing role types was to help reduce the complexity of an inter-connected system (Ferdows 1997). Ferdows also indicated the upgrading path from one type of role to another and formulated the relationship and strategy road map of those role types very clearly. However, after categorising the role types in a business ecosystem, Iansiti and Levien did not touch on the transformation among different roles, even though they thought the roles could be changed over time. Based on Iansiti and Levien's four types of role, Iyer and his colleagues proposed three roles: the hub (a firm with a disproportionately high number of links) to select strategies of keystone, dominator or niche; broker (a firm that creates a connection between two sets of firms); and bridge (a link critical to the overall connectedness within the network) (Iyer et al. 2006).

After introducing role types and strategies, Iansiti and Levien also studied how firms competed with one another based on three founda-tions: architecture, integration and market management. Architecture (standards, platform and architecture) describes how firms draw bounda-ries among technologies, products and organisations, which is normally as an interface for partners' interaction and value adding. Platform strategy was specifically highlighted as it could demark the boundary between competition and cooperation, and increase the players' performance (Bannerman & Zhu 2008). Integration defines how organ-isations collaborate across these boundaries (architecture) in order to share capabilities and technological components. Market management shapes how organisations complete transactions across these boundaries (architecture) in the context of complex and dynamic markets (Iansiti & Levien 2004b).

Furthermore, Iansiti and Levien also tested the health of a business ecosystem by using three dimensions: productivity, robustness and niche creation (Iansiti & Levien 2004a, 2004b). Productivity is the network's ability to consistently transform technology and other raw materials of innovation into lower cost and into new products. Three measures to test productivity are factor productivity, change in productivity over time and delivery of innovations. Robustness indicates that a business ecosystem should be capable of surviving disruptions such as unfore-seen technological change. There are five measures to check robustness: survival rates, persistence of ecosystem structure, predictability, limited obsolescence and continuity of using experience and cases. Niche crea-tion demonstrates the ability to maintain the growth of firm variety, product and technical variety. These three dimensions expanded the criteria to measure a system's health. For example, Christopher thought supply-chain managers should test supply-chain health with four aspects:

quickness of response to market, reliability, resilience to external shock and the relationship among different partners (Christopher 2005). The detailed content of 'productivity' and 'robustness' extended the scope of 'responsibility', 'reliability', and 'resilience'. 'Niche creation' also highlighted the growth of firm and product variety rather than simply the 'relationship' between suppliers and customers (Iansiti & Levien 2004b; Christopher 2005).

Later in 2006, Iansiti also proposed a framework by synthesising ecosystem structure, health measurement methods and performance to explore IT ecosystems, by which the network of organisations drives the delivery of information technology products and services (Iansiti & Richards 2006). The structure of this IT ecosystem is increasingly open and competitive, with a growing number of application providers supporting multiple platforms (Iansiti & Levien 2004b).

In summary, Iansiti and Levien distinguished the role types which matched the core ideas of Ferdows's strategic roles of foreign factories within an international manufacturing network. However, Iansiti and Levien did not study the transformation between different roles. Iansiti also proposed the health measurement criteria to be productivity, robustness and niche creation. However, this study provided a descriptive framework rather than practical measurement methods. Furthermore, Iansiti and Levien mostly focused on firms' strategies in a business ecosystem rather than the business ecosystem itself. More work is needed on business ecosystem construct studies.

Adner, Kapoor and his colleagues 2006, 2010, 2012, 2013 and their edited book 2013

Adner and his colleagues studied a business ecosystem via a stable structure called structure of technology independence (Adner 2006; Adner & Kapoor 2010; Kapoor & Lee 2013). In this structure there are focal firms who provided the platform and their component suppliers, customers as well as the complementors. They addressed the functions of complementors very much since the complementors facilitate the focal firms' product commercialisation and add value to the focal firms' products. Thus, the complementors will enable the customers to better use the focal firms' products (Adner & Kapoor 2010; Adner 2012).

They further test the relationship between focal firm and component suppliers and complementors. They have found that greater technological challenges in the component side will increase the performance advantage of focal firms. However, greater technological challenges in complements will decrease the focal firms' performance advantage.

Furthermore, along the life-cycle development, focal firm tended to integrate more partners in order to reduce the partners' behaviour uncertainty (Adner & Kapoor 2010).

Besides the structure study, in 2006 Adner also suggested that a firm should consider three fundamental risks in an innovation ecosystem in terms of its structure of technology interdependence (Adner 2006). They were initiative risks: the familiar uncertainties of managing a project; interdependent risks: the uncertainties of coordinating with complementary innovators; and integration risks: the uncertainties presented by the adoption process across the value chain. When a firm has set performance expectations and target markets and taken a role of keystone, niche player or dominator (Iansiti & Levien 2004b), they should carefully consider the three fundamental risks before modifying their performance expectations and determining where, when and how to compete.

Later, in 2013, Kapoor and Lee also studied the focal firm's technology investment based on that structure of technology interdependence. They have found that firms owning a bigger scope of alliance with their complementors intended to invest more in technologies than those integrated complemenotrs or were very independent from complementors (Kapoor & Lee 2013).

However, their studies and the book they edited in *Advances in Strategic Management* (Brusoni & Prencipe 2013; Kapoor 2013; Mäkinen & Dedehayir 2013; West & Wood 2013; Li & Garnsey 2013) mainly focus on a stable and very small structure of a business ecosystem, addressing the complex relationship between focal firms and complementors, which cannot well explain the dynamic nature and give a comprehensive picture of a business ecosystem composed of different levels of stakeholders. For example, the structure at the early stage of a business ecosystem was very fragmented and was gradually being a mature industrial system (Moore 1993). In that way, Adner and their colleagues' framework was only testing the mature stage of a business ecosystem, which was more like the supply-chain management.

Peltoniemi and his colleagues in 2004, 2006

As the contents of Peltoniemi and his colleagues' papers are partly overlapped (Peltoniemi & Vuori 2004; Peltoniemi 2004; Peltoniemi et al. 2005; Peltoniemi 2005; Peltoniemi 2006), three of their papers are selected to demonstrate the key ideas on the business ecosystem study.

In 2004, in order to discuss the drivers and significant potentials of the business ecosystem study, Peltoniemi proposed a comparison among the concepts of cluster, value network and business ecosystem (Peltoniemi 2004). This chapter also supports Moore's view (Moore 1996) as to why to study business ecosystems. 'Cluster' refers to geographic concentration, and the competition inside a cluster will be very fierce, while 'value network' indicates a strictly cooperative structure. Both competition and cooperation exist in a business ecosystem in forms other than clusters and value networks. As a result, knowledge is more easily transferred within the business ecosystem than it is within a value network or cluster. Three different innovation strategies are to be adopted by the three different types of network. Firms in a cluster will imitate other players' innovations; firms in a value network will act conservatively to be compatible with their partners; firms in a business ecosystem will be interdependent and will exploit opportunities to achieve co-evolution.

In order to distinguish the business ecosystem study from the study of other analogies of biological ecosystems, Peltoniemi and Vuori (2004) reviewed industrial ecosystems, economy as an ecosystem, digital business ecosystems and social ecosystems. This work helped build up the research boundaries for the business ecosystem study. Their work also explored complexity theory within the business ecosystem context, as the business ecosystem reflected the complexity among different roles (Moore 1996). Four features were identified to describe the business ecosystem from a complexity system perspective: self-organisation; emerging potential for creativity; evolution; and adaptation (Peltoniemi & Vuori 2004).

In 2006 Peltoniemi integrated the 'complexity system' model with an evolutionary economics view to propose a preliminary theoretical framework of the business ecosystem. In this new framework, three interactions of variation, selection and development were added as well as conscious choice, limited knowledge, interconnectedness and feedback. Thus the business ecosystem could be regarded as different organisations which are interconnected through competition and cooperation (which can be present simultaneously). Feedback loops trigger a dynamic future while organisations display different behaviours based on conscious choices (Peltoniemi 2006). All of these loops demonstrated the view of complexity and evolution.

In summary, Peltoniemi's contribution distinguished the difference between the business ecosystem and its ecosystem-related area. He also compared the business ecosystem with value networks and clusters from the perspectives of cooperation and competition. However, he simply combined clusters with competition orientation and value networks with cooperation orientation. Actually, competition and cooperation both exist in these two networks (Porter 1990; Passiante & Secundo 2002; Bell 2005). Furthermore, Peltoniemi's theories developed a descriptive framework of a business ecosystem from a complex system and an evolutionary economics perspective. Peltoniemi did not explore the detailed constructs of a business ecosystem and its growth.

Den Hartigh and his colleagues in 2004, 2006, 2007

In 2004 Den Hartigh and his colleagues thought the business ecosystem boundaries should be defined by the positions of the various role players. For example, a specific business ecosystem boundary could be identified as the firm owning the core technology together with its relevant partners (Den Hartigh & Van Asseldonk 2004). Thus the boundary would vary from firm to firm. Furthermore, they integrated the four types of role and their strategies which had been suggested by Iansiti and Levien into three roles: shaper, adapter and reserving the right to play. The shaper offers the platform or core technology and encourages other partners to participate. It could also perform as the keystone, dominator or landlord as Iansiti and Levien mentioned (Iansiti & Levien 2004b). Firms choosing adapter strategies would provide products that are complementary to or compatible with the shaper's core technology. In this strategy they argued that niche players cannot explain all the adapter's behaviour as adapters sometimes also directly influence the shapers' strategy. The third role involved those who acted as opportunists and did all the necessary business.

In 2006 Den Hartigh and his colleagues also further developed health measurement methods from the firm's perspective. Previously, Iansiti and Levien proposed health measurement on the business ecosystem level (productivity, robustness and niche creation). In order to improve the measurement for practical use, Dean Hartigh and his colleagues mapped and transferred health measurement criteria from meso-level (Iansiti & Levien 2002, 2004b) into firm level. Thus this chapter described a health measurement instrument for industrial managers' practical use which included benchmarking ecosystem performance, assisting in the partner selection process and a possible compass for ecosystem governance (Den Hartigh et al. 2006).

In 2007 Den Hartigh and his colleagues provided an overview of business research and defined the four core parts of the business ecosystem as the firms, the network, performance and governance. These four parts were generated by comparing social network theory, biological ecosystem theory, complex adaptive system theory and business ecosystem theory. They addressed the firms' role type and strategies, the network's structure and dynamic, performance both at the firm and network levels and governance at the network level. This integrated framework extended the traditional strategic management studies which were at the level of core products, service and network.

In summary, Den Harigh's work clarified the business ecosystem boundary as a core firm with technology and its relevant partners. Then he further developed the business ecosystem based on Iansiti and Levien findings: on the one hand, he integrated four roles' types into three types as the 'shaper', 'adapter' and 'reserving the right to play' with a more practical orientation; on the other hand, he brought in health measurement instruments which enable productivity, robustness and niche creation to be measured. Finally, he also proposed an integrated framework to describe the business ecosystem with firms, network, performance and governance. However, this framework was still on a descriptive level and did not explore business ecosystem constructs.

Other contributors after 2002

Besides, the Iansiti and Levien (2004) work in platform-based ecosystem, Gawer and Cusumano also mentioned there were two kind of platforms to organise the business ecosystem. The first is the internal platform, which was the company-specific platform and could be used among connected supply-chain partners, while the second one is the external and industry-specific platform like the Android-based operating system platform which was open and could be used by all ecosystem partners. The key platform should be the industry platform which could facilitate the complement services and products (Gawer & Cusumano 2013). Furthermore, based on that industry platform, more and more independent software vendor (ISV) participation will add more value to the platform ecosystem. In detail, the ISV could perform better if their product IP were protected or they had downstream complementary capabilities to organise the customers and relevant partners (Ceccagnoli et al. 2012).

Manning and Thorne applied Moore's business ecosystem ideas (Moore 1996) to a customer-driven situation and proposed six steps to build demand-driven ecosystems, including understanding customer and mapping ecosystem, leverage partners, building adaptive strategies, using information technology, building ecosystem capabilities and synchronising supply and demand (Manning et al. 2002).

Later some authors studied how to describe business ecosystem activities in a comprehensive picture and then suggested different governance structures for a business ecosystem: 1) Iyer and his colleagues (Iyer et al.

2006) suggested five factors to examine the stability of network alliances: average number of partners; degree of linkage; cluster coefficient; path length; and network density. They applied this structure to a software sector ecosystem and found it has a small-world structure: very efficient in moving information, innovations and other resources through the ecosystem. 2) Adomavicius and his colleagues introduced the idea that a technology ecosystem could be briefly defined as a focal technology within the given context. The context consisted of social and government forces, technical forces and economic forces. Roles of technology are divided into Component, Products, Support and Infrastructure which were also the key factors determining technology patterns and their evolution in the business ecosystem (Adomavicius et al. 2006). 3) A 'Dynamic agent-based ecosystem model' framework was proposed for modeling business ecosystems, containing sets of parameters like environment size, amount of servicer offers and number of agents and so on, as a synthesis of ideas from natural ecosystems and multi-agent systems (Marín et al. 2007). 4) Chang and Uden proposed a three-way framework for governing an e-learning ecosystem which included organisational decision-making structure, operational and technical support processes and communications and relational mechanisms (Chang & Uden 2008).

Besides these, other authors have also studied business ecosystems at strategic levels: 1) Gueguen proposed the business ecosystem as the dynamic of collective strategies and first identified ecosystem competences/capabilities as a 'Collective combination of different shareable resources and competencies from different key players' (Gueguen et al. 2006). 2) In 2009, Li (2009) proposed that a business ecosystem has three major characteristics: symbiosis; platform; and co-evolution. The 'symbiosis' indicates partners of a loose network with synergic and systematic cooperation; the 'platform' was the service, tools or technologies that other partners in the ecosystem can use to enhance their own performance, meaning the value of a business ecosystem shifts from product to network; the 'co-evolution' demonstrated different roles such as keystone, niche and dominator which cooperated and were complementary to one another with the purpose of mutual benefit. 3) William and De Meyer suggested six strategic ways to unlock ecosystem advantages such as pinpointing where value is created, defining an architecture of differentiated partners' roles, stimulating complementary partners' investments, reducing the transaction costs, facilitating joint learning across the network, and engineering effective ways to capture profit (Williamson & De Meyer 2012); 4) Power and Jerjian proposed 12 strategies to run a web-based business ecosystem as service communities (Power & Jerjian 2001; Desai et al. 2007).

Conclusion of the business ecosystem studies

In summary, the evolutionary path of business ecosystem theory development is presented as Figure 3.5. There are three streams of the business ecosystem study: life cycle, system structure and performance.

In the first stream, 'Life cycle study' shows how a business ecosystem evolves from phase of birth, expansion, authorities towards renewal (Moore 1993). In 2002, a demand-driven customer ecosystem was proposed as well as its building process, including: understanding customers and mapping the ecosystem; leveraging partners; building adaptive strategies; using information technology; building ecosystem capabilities; and synchronising supply and demand (Manning et al. 2002).

In the second stream, 'system structure', Moore first proposed seven business issues of developing a business ecosystem: the customer, market, offer, business process, organisational architecture, stakeholder and societal value, and government policy (Moore 1996). Following this work, many scholars continued to develop other relevant governance structure models: structure from evolutionary economics and complex system perspective, integration of market dynamic and firm conduct, small network structure, an integrated structure of firm, network, performance and governance and e-learning governance framework of structure, process and relational mechanisms (Den Hartigh & Van Asseldonk 2004; Peltoniemi & Vuori 2004; Peltoniemi 2006; Adomavicius et al. 2006; Iyer et al. 2006; Anggraeni et al. 2007; Chang & Uden 2008; Adner & Kapoor 2010; Kapoor & Lee 2013). Finally, Moore also proposed the output of those structures as eight public 'goods' (benefits) (Moore 2006).

Besides the whole system structure, many scholars also paid attention to role identification (Moore 1993; Iansiti & Levien 2004b; Den Hartigh & Van Asseldonk 2004; Iyer et al. 2006). Moore presented the business ecosystem definition of different levels of organisations. Iansiti and Levien further gave details of role types as the keystone, niche, dominator and hub landlord (Iansiti & Levien 2002). Those four roles were further integrated into three roles with more functions as the shaper, adapter and opportunist (Den Hartigh & Van Asseldonk 2004). In 2006, Iyer and his colleagues also proposed three types of roles, bridge, hub and broker (Iyer et al. 2006). Adner also addressed the focal firm and complementary relationship(Adner & Kapoor 2010; Adner 2012; Kapoor & Lee 2013).

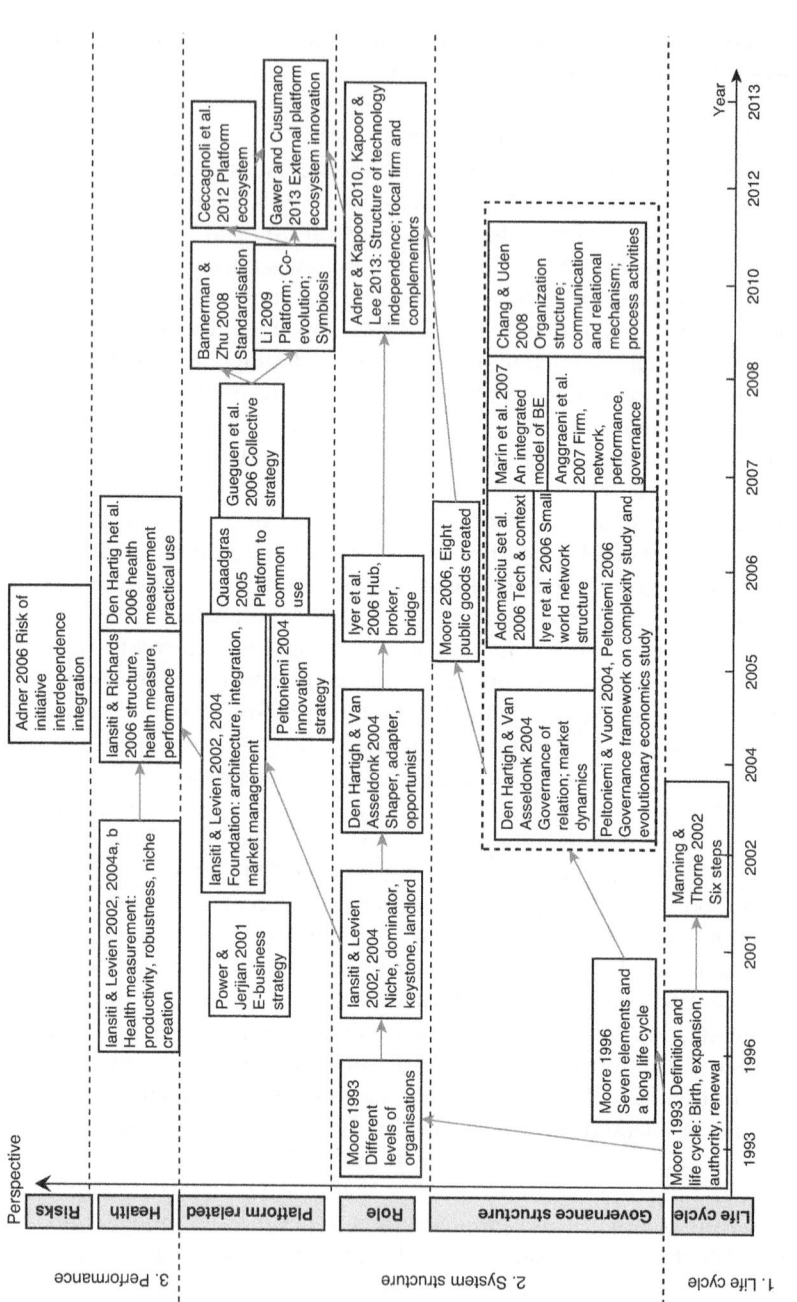

Figure 3.5 Road map of business ecosystem literature since 1993

Furthermore, the platform and its related issues were highlighted as they could make partners' interaction more easy and efficient (Power & Jerjian 2001; Iansiti & Levien 2002; Quaadgras 2005; Peltoniemi 2004; Gueguen et al. 2006; Bannerman & Zhu 2008; Moore 1996): ICT web as a platform to connect firms; technology platform and its architecture to connect partners with complementary components; standardisation to improve industry productivity (Ceccagnoli et al. 2012; Gawer & Cusumano 2013).

Besides studying the system and its lifecycle, the third stream focused on the performance of a business ecosystem. The measurement methods of ecosystem health were proposed as productivity, robustness and niche creation (Iansiti & Levien 2002; Iansiti & Levien 2004b; Iansiti & Richards 2006; Den Hartigh et al. 2006). Risk management was also proposed regarding the performance of the business ecosystem. There was risk happening around initiative, interdependence and integration activities within a business ecosystem.

3.3.3   Ecosystem study on different perspectives

The ecosystem study contained different branches such as business ecosystem, industrial ecosystem, social ecosystem, digital business ecosystem and economy as an ecosystem (Peltoniemi & Vuori 2004). This section will explore other ecosystem studies.

In Henderson's dictionary of biological terms, ecosystem is defined as a community of different species interdependent on one another together with their non-living environment, which was relatively self-contained in terms of energy flow, and was distinct from neighbouring communities. Different types of ecosystem were defined by the collection of organisms found within them. Ecology is the study of the inter-relationships between organisms and their environment (Lawrence 1990). The idea of a biological ecosystem was first proposed by Darwin, who thought living organisms were co-evolving with their non-living environment by natural selection (Darwin 1859). This idea opened up the possibility of studying the natural world as a whole system rather than as a collection of unconnected items.

The biological ecosystem idea was also applied to different research areas. Table 3.6 shows the evolution of applications of ecosystem ideas to other areas such as originality, social, business and technology. Many academic researchers have applied ecosystem ideas to social studies (Bronfenbrenner 1977; Bronfenbrenner 1994; Mitleton-Kelly 2003). The ecology of human development studies the relationship

Table 3.6 Comparison of the biology ecosystem – related theories

Perspective	Theory	Time first proposed	Originating author	Content related to the ecology	Key conclusion word
Originality	Darwinism, biological ecosystem	1859	Darwin	Natural selection	Survival
	General systems theory	1936	Ludwig von Bertalanffy	Open system; highlight the importance of environment; flexible institutional structure ; dynamic interactions	Open with environment
Social orientation	Ecological model of human development	1977	Urie Bronfenbrenner	Understand human development by five sub-ecosystems: microsystem, mesosystem, exosystem, macrosystem, chronosystem	Five sub-ecosystems for human ecological model
	Social ecosystem	2003	Mitleton-Kelly	Each organisation influences and is influenced by social ecosystem organisation	Interdependence, co-evolution
Organisational orientation	Organisational ecology	1977	Hannan & Freeman; Van Witteloostuijn	Organisational ecology seeks to understand the dynamics of change that take place within organisational populations	Organisational founding, change and mortality with organisational environment
Business orientation	Industrial ecosystem	1989	Frosch & Gallopoulos	Environment protection when manufacturing	Environment friendly when industrial operation
	Economy as an ecosystem	1990	Rothschild	Economy development not only shaped by its own genes but also defined by its relationships	Interaction
	Business ecosystem	1993/ 2004	James F. Moore/ Iansiti & Levien	Dynamic interactions; consumers are part of the whole system; keystone; niche; share value	Co-evolution; system competition
Technology orientation	Digital business ecosystem,	2002	Nachira et al.	Adoption of Internet-based technology for business, supported by pervasive software environment	Internet-based software environment, evolutionary and self-organisation

Source: Partly adapted from Peltoniemi & Vuori (2004).

between a type of organism and the changing environment in which it lives (Bronfenbrenner 1977). Bronfenbrenner divided the ecological model of human development into five sub-ecosystems in order to understand and guide human growth. They were microsystem, meso-system, exosystem, macrosystem and chronosystem. A microsystem is the pattern of activities and social relationships of a given person in a set of individuals. A mesosystem is a system of microsystems which comprises the linkages and processes taking place among two or more sets of individuals. An exosystem comprises the linkages and proc-esses taking place among two or more sets of individuals, at least one of which does not contain the original person. A macrosystem consists of the micro-, meso- and exosystem characteristics of a given culture or subculture. Chronosystems include the environment and its changes to a macrosystem over the lifetime of the individuals (Bronfenbrenner 1994). In 2003, Mitleton-Kelly proposed the concept of social ecosystem where organisations influenced and were influenced by all other related organisations such as business, consumers, suppliers and legal institu-tions (Mitleton-Kelly 2003). In conclusion, this view of ecosystem study was to apply Darwin's natural selection implications to social systems.

Organisational ecology was introduced by Hannan and Freeman who addressed the organisation–environment relationship as an alterna-tive to the dominant adaptation (Hannan & Freeman 1977). It was also developed on the solid foundation of a well-communicated and well-established theoretical core – an organisational translation and exten-sion of Darwinian biology (Van Witteloostuijn 2000). Organisational ecology aimed to understand the dynamic change that took place within organisational populations, as well as to understand how organisational characteristics, ecological determinants and macro-environmental conditions affected the rates of organisational founding, change and mortality (Messallam 1998).

Besides broadly studying social or organisational ecosystems, some researchers began to use the concept in the business world (Frosch & Gallopoulos 1989; Rothschild 1992; Moore 1993; Iansiti & Levien 2004b). Frosch and Gallopoulos (1989) presented an environmentally friendly manufacturing concept as the industrial ecosystem in which materials were well used to reduce harm to the future environment. In 1992, based on environmentally friendly thinking, Rothschild also applied the ecosystem concept to run firms' business which was not only shaped by its own genes but also defined by its relationships (Rothschild 1992). To make the business orientation more detailed and operable, Moore and Iansiti developed business ecosystem theories from the perspectives of business activities, life cycle, role types, key strategies and evolution (Moore 1993; Iansiti & Levien 2004b).

As ICT technology was one of the major contributors to economic growth and economic efficiency (Martin 2010), a digital business ecosystem was launched as a European Union programme in 2002 to support large numbers of SMEs to compete with larger software companies (Nachira 2002). However, this kind of business ecosystem did not have many features in common with the implementation stage of an original business ecosystem (Peltoniemi & Vuori 2004). It focused on Internet-based technology with a digital environment such as software components, applications, service and knowledge (Nachira et al. 2007).

In summary, this section has developed two findings: first, it briefly reviewed typical ecosystem studies with different perspectives along a timeline. Table 3.6 illustrated the history of the idea of biological ecosystems as applied to the study of social and management phenomena during the last century. From general systems theory, organisational ecology and human development ecology, to business ecosystem, theories address the interaction between organisms and the environment in which they live and grow. Secondly, by reviewing the different branches of ecosystem study, this section also clarified the research boundary of this study on business ecosystems. The business ecosystem highlights the firm's co-evolution with the business environment. Faced by rapid change in today's industry era, the business ecosystem gives a new perspective from which to view companies. Managers are able to find totally new things that they had never found before (Moore 1993).

3.4   Identification of theoretical and practical gaps in the business ecosystem literature

From the business ecosystem literature, the life-cycle study still had much space for improvement. On the one hand, the BELC study was still at its early stage. It focused on the level of the ecosystem growing rather than the firms' activities along the BELC. On the other hand, Chapter 2 developed the industry challenges: 'in the recent dynamic environment, how could firms quickly capture and commercialise ideas, then maintain the process sustainably, and finally successfully nurture a new industry?' As a result, the detailed transformation activities between different life-cycle phases should be explored further in order to understand how firms evolve along the BELC as Table 3.7.

Besides the life cycle and its nurturing process study, business ecosystem theories have not been explored comprehensively since 1993.

Table 3.7 The key areas for network study

Key areas	Sub-dimensions	Authors	Relevant theories
Life cycle	Phases	(Rink & Swan 1979); (Peltoniemi 2011); (Moore 1996)	Product life cycle; Industry life cycle
Nurturing Process along lifecycle	Nurturing process Strategy	Mills et al. 1995; Polli & Cook 1969; Hayes & Wheelwright 1979; Garnsey 1998; Klepper 1997	Operation management, GMVN, open innovation, firm growth
Constructive elements	Structure Infrastructure	Hayes & Wheelwright 1984; Shi & Gregory 1998; Harland 1996; Zhang et al. 2007	Operation management, IMN, supply chain, GEN
Configuration pattern	The integrated pattern based on constructs	Adapted from Zhang et al. 2007, Srai & Gregory 2008, Shi & Gregory 1998	Operation management, IMN, supply chain, GMVN

Various different areas were explored such as system structure (roles, governance structure, platform) and performance (health and risks) in Table 3.7. However, due to the limitation of published papers, the study on those areas was also not systematically organised. So it is necessary to conduct a systematic study of previous manufacturing systems, and to suggest areas for future research on the business ecosystem as a whole.

The constructs of a business ecosystem should be explored further as a system could be understood through exploring its constructive elements (Von Bertalanffy 1969). In 1984, Hayes and Wheelwright (1984) highlighted a constructs study with the framework of 'structure–infrastructure' model, since constructive elements had a big impact on system manufacturing strategy. This model was popular and continuously adopted by scholars at other manufacturing system levels such as the intra-firm level (Shi & Gregory 1998) and inter-firm supply chain level (Harland 1996), global engineering network, and global supply network level (Zhang et al. 2007; Srai & Gregory 2008). As a result, in order to keep a close relationship with classical theories such as the manufacturing system and supply chain management theory, the framework of 'structure–infrastructure' is adopted to deconstruct a business ecosystem.

However, according to general systems theory, the individual constructive elements and nurturing process could not provide a complete picture of a system, because there are further system activities of coordination between constructive elements and processes (Von Bertalanffy 1950). The integration of the constructive elements and processes of each system delivers various configuration patterns which demonstrate the typical manufacturing strategy. Hayes and Wheelwright used two elements, process and product, to categorise different patterns of manufacturing system such as project-based, job flow, batch, line flow and machine-paced flow (Hayes & Wheelwright 1979; Hayes & Wheelwright 1984). Shi and Gregory (1998) extended the configuration pattern concept to network level with geographic dispersion and manufacturing coordination which assumed that an international manufacturing network demonstrated the typical way of organising a manufacturing network. In 2007, Zhang's 3C model (Context, Configuration and Capability) described the configuration of a global engineering network. In 2008, Srai's four dimensions also configured a supply network. Both these studies, to some extent, followed a structure–infrastructure perspective by proposing the configuration pattern (Zhang et al. 2007; Srai & Gregory 2008). These key research areas are listed in Table 3.7 with sub-dimensions, key authors and relevant theories.

By investigating both the industry and the literature, key research gaps are identified for study of the business ecosystem as Figure 3.6. They are: 1) business ecosystem life cycle; 2) constructive elements that act as the fundamental skeleton of a business ecosystem; 3) configuration patterns as an integrated model to demonstrate typical ways of nurturing a business ecosystem; 4) nurturing process along the BELC in order to reach the strategic objectives.

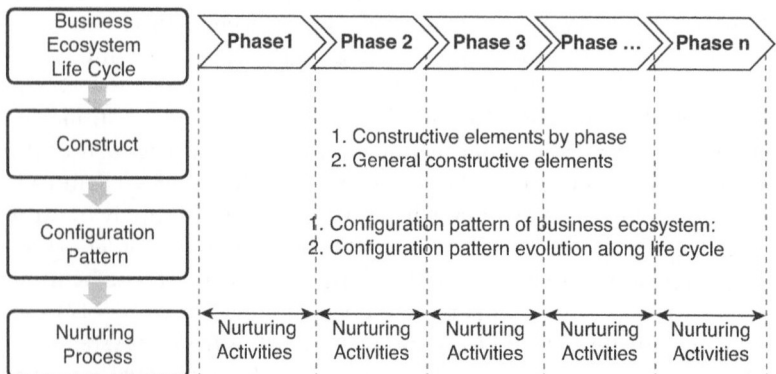

Figure 3.6   Research gaps of business ecosystem

3.5 Conclusion

In order to identify research gaps for the business ecosystem studies, this chapter conducts a systematic review on the previous manufacturing systems-related theories.

Primarily, manufacturing system theories on firm level, intra-firm level, inter-firm level and internationalised inter-firm level were reviewed with reference to two industrial challenges: uncertainty and interoperability. However, none of these theories can well tackle the problems of these two industry challenges. As a result, due to the nature of co-evolution, the business ecosystem theory is introduced as a potential area to solve industry challenges.

Secondarily this chapter provided a literature review on business ecosystems, with a historical perspective in terms of its definition, key contents and other relevant ecosystem studies.

Finally, as the business ecosystem theory is still at an early stage, more in-depth research should be conducted by addressing four research gaps of the life cycle, nurturing process, constructs and configuration patterns. The first research gap, 'business ecosystem life cycle', is to explore the general evolutionary stages of a business ecosystem. The second research gap is to identify the key elements with which a business ecosystem is constructed. The third research gap is to identify integrated models as the configuration patterns on how a firm nurtures a business ecosystem. The final research gap is to describe the firms' nurturing activities under the general context of the business ecosystem life cycle.

4
Research Design

4.1 Introduction

This chapter has two aims:

- to discuss the key research framework and research questions;
- to develop the appropriate research methodology for this research including the choice of research methodology, procedure, data collection and analysis. All of those methods will ensure the rigorous research with internal validity, construct validity, external validity and reliability. (Gibbert et al. 2008)

4.2 Research framework and objectives

4.2.1 Research questions

Learning from the research gaps of the literature review, four sub research questions are identified as:

1) **What is the life cycle of a business ecosystem?** How many phases does the business ecosystem life cycle (BELC) have? What are their phase-ending statuses?
2) **What are the main constructs of a business ecosystem?** What are the key constructive elements of a business ecosystem in terms of life-cycle phases; what are the general constructive elements of a business ecosystem?
3) **What are the configuration patterns of a business ecosystem?** What are the key constructive elements that can be used to categorise different configuration patterns of business ecosystems? What are those patterns? Are there any typical patterns in the evolution model along the nurturing process?

4) **How to nurture a business ecosystem along the life cycle?** What are the different nurturing processes in terms of life-cycle phases?

By synthesising those four sub-questions, the main research question will be identified as:

4.2.2 How to nurture a business ecosystem from firm perspectives?

From these four questions, this study aims to achieve the following two research objectives:

- To achieve the theory building on the BELC and its constructs, configuration pattern and nurturing process.
- To provide the theoretical and practical implications from the business ecosystem research.

4.2.3 A general framework

A business ecosystem is an economic community where firms co-evolve with their business environments (Moore 1993). The business ecosystem helps firms realise the importance of interaction with their different levels of partners in order to capture dynamic business opportunities. From the literature review, current theories are focusing on the BELC, the ecosystem structure and the ecosystem performance; however, there is still much space left for further exploration.

In order to answer these four sub-questions, a framework should be developed in order to investigate this question. Figure 4.1 presents the development of the research process framework based on current theories.

The research process framework aims to link industrial challenges, theoretical background, research questions and outcomes which could guide the case studies and data analysis in the right direction: the left part – preliminary research – presents research background and practical review. The middle part – literature review – studies manufacturing system theories and the business ecosystem. Based on the comparison between the practical review and literature review, the research gaps of the business ecosystem are identified as the life-cycle and nurturing process, constructs and configuration pattern. As a result, the outcomes of the right part will be generated in detail.

Following the research process framework, the conceptual framework is proposed as Figure 4.2 in order to connect those outcomes as a logic system to achieve the research with internal validity (Gibbert et al. 2008; Lin & Zhou 2011).

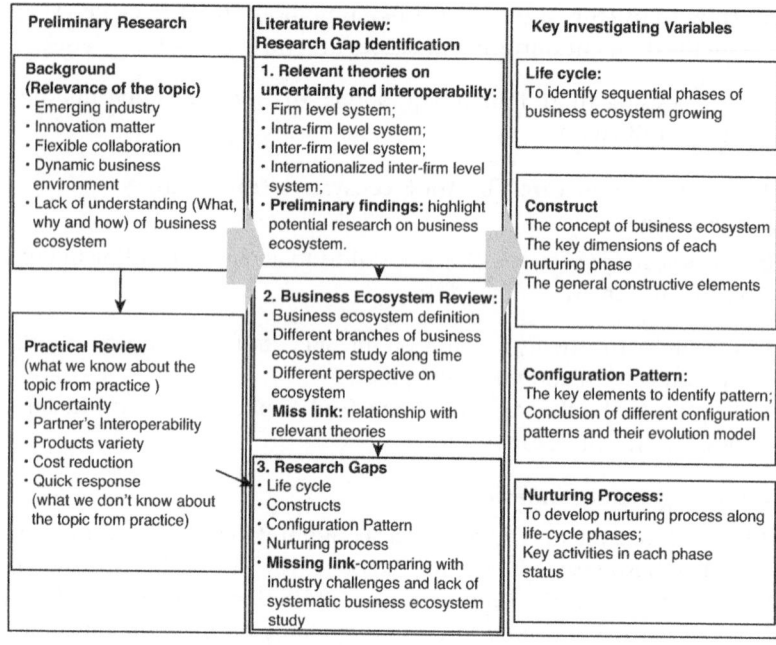

Figure 4.1   Research process framework

The life cycle of a business ecosystem

The life cycle part studies a business ecosystem evolution which extends the scope of the product life cycle and the industry life cycle by highlighting the co-evolution issues between firms and their business environment, emphasising the inter-industries study and the idea of sustainability of the life cycle. A business ecosystem at different phases of its life cycle will generate different contexts and implications for firms' development. As a result, firms at different phases of the BELC will need different nurturing processes.

The constructive elements of a business ecosystem

In order to study the nurturing process activities at a fundamental level, the constructive elements should also be explored deeply, because the constructive elements are the key and most crucial dimensions to represent the network strategies (Shi & Gregory 1998) and also have a big impact on the nurturing process. As Moore's framework is not of systematic study (Moore 1996), the classical framework of 'structure-infrastructure' is adopted to study a business ecosystem at a fundamental level as well as to maintain a close relationship with traditional theories (Hayes & Wheelwright 1984; Shi & Gregory 1998; Harland 1996). Furthermore, the constructive elements also have different impacts on each life-cycle phase as the business ecosystem evolves. A comparison between the constructive elements across phases should also be conducted.

The configuration pattern of a business ecosystem

The configuration pattern is usually identified by the elements reflecting the constructive elements and real industry requirements which provides a better way of concisely representing a complex organisation and forming an integrated view (Shi & Gregory 1998; Zhang et al. 2007). The configuration patterns are regarded as the categories of systems' format with specific objectives and operations. So the real cases can help develop classifications of the configuration patterns with different objectives and operations. Besides static patterns, this research also explores how the business ecosystem evolves with different patterns along the life cycle.

The nurturing process along the BELC

The nurturing process is to deeply explore the transformation activities along each phase of the BELC. In 1996, Moore proposed a general picture of how a firm nurtured its ecosystem with four sequential steps including birth, expansion, authority and renewal (Moore 1996). However, Moore did not touch on the detail of the transformation process of a firm nurturing the business ecosystem along its life-cycle phases. Furthermore, the new emerging industrial phenomenon of uncertainty and interoperability also challenges Moore's four phases. As a result, the detail transformation activities need further exploration along the BELC.

In conclusion, Figure 4.2 describes the conceptual framework of the business ecosystem. The BELC will be divided into sequential phases (to make it clearer, each phase-ending status will be identified too). The other parts contain the nurturing process, constructive elements and configuration pattern along the life-cycle phases. Thus, different phases will have different nurturing activities. The constructive elements will have different impacts on each phase of the life cycle. Regardless of the life-cycle perspective, the business ecosystem itself also has its general constructive elements. The constructive elements and nurturing process will help define different configuration patterns of a business ecosystem from firm perspectives. The configuration pattern evolution models along the life cycle are also to be explored.

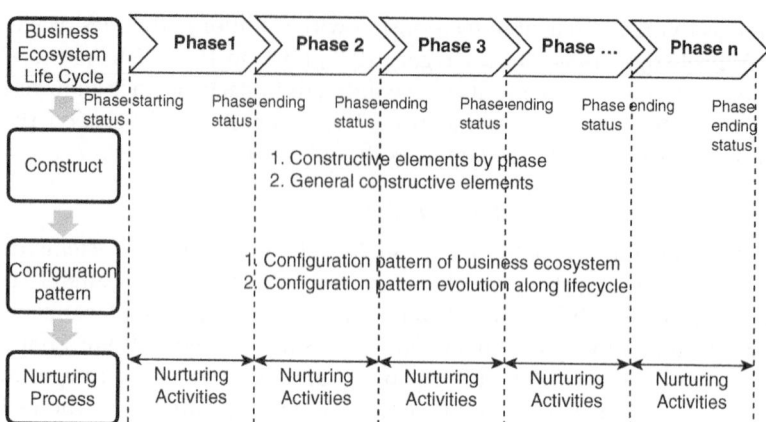

Figure 4.2   Conceptual framework of the business ecosystem

4.3   Overview of research methodology

4.3.1   Case studies method

The selection of appropriate research methods is significant because the research approach influences the quality of management research (Easterby-Smith 1997). This research will conduct a case study with a theory-building nature (Eisenhardt 1989; Yin 2008) in order to systematically explore the business ecosystem theories. As the research question is 'how to nurture a business ecosystem from a firm's perspective', it demands a good understanding of the process of managing the business ecosystem for emerging ideas and technology. The unit of analysis could be the sequential projects which are implemented by a firm owning its ecosystem. The firm aims to nurture an idea or technology by running sequential projects with partners in the business ecosystem. So the detailed nurturing process of the business ecosystem could be figured out by deeply exploring a series of projects between the firm and its ecosystem partners.

As a result, in the proposed research good cases will be the key linkage between practice and literature. Suitable case firms will be chosen using the following seven criteria:

1) The case companies already have their own business ecosystems or are willing to set up new business ecosystems.

2) The case companies, as chip providers, are central to their partners' network.
3) Each case's business ecosystem should have a different development status. The first case is at the sustaining status as its business ecosystem gradually grows; the second one is at the transforming status as its business ecosystem experiences the change; the third one is at the initialising status as its business ecosystem is building from the beginning.
4) The companies are involved in the high technology semiconductor-related industry. The first company is mainly from the mobile phone industry; the second company is mainly from the PC industry; the third company is mainly from the home device industry.
5) The companies have the solution platforms for their partners' participation and contributions.
6) The case companies have sequential projects to demonstrate how they build up their business ecosystem. Those sequential projects within each case could together demonstrate the whole nurturing process along the business ecosystem life cycle.
7) The companies have continuous innovation for future products in the mobile computing industry.

Table 4.1 shows the difference among these three cases according to the seven criteria which aim to find three typical cases with different

Table 4.1 Selective criteria for main cases' studies

Cases Criteria	ARM	Intel	MTK
(1) Business ecosystem	Very open	Less open	Less open
(2) Central firm	Yes	Yes	Yes
(3) Ecosystem development status	Sustaining	Transforming	Initialising
(4) Firms' background	Mobile as the main product	PC as the main product	Home device as the main product
(5) Solution platform	Highly open chip platform	Less open chip platform	Highly integrated chip platform
(6) Sequential projects to demonstrate whole nurturing process	Yes	Yes	Yes
(7) Future product	Mobile computing	Mobile computing	Mobile computing

priorities. The first four criteria are to select specific companies as the main cases while the last three criteria are to help identify the relevant projects within the selected main cases.

Following those criteria, ARM, Intel and MTK were selected. ARM is the IP provider (fundamental basis of chips). It started to build its ecosystem from the early 1990s and set up a department of connected community in 2003 which was just for enhancing ecosystem development. ARM's ecosystem was very successful as 98 per cent of mobile phones were based on ARM's platform with more than 500 partners in its ecosystem. ARM's ecosystem kept growing gradually. Intel also started to build an ecosystem very early when they focused on the PC industry. They first developed the public interface with open code to connect the chip set from every partner inside the business ecosystem. After dominating the PC industry, Intel aimed to copy its PC model into the mobile computing industry, but without success. Then Intel was at the transforming status to reconsider its ecosystem strategy. The third main case is MTK, which is a unique company with well-accepted product solutions. They provide a turnkey (one-chip solution) model chip which integrates all the chips with essential functions for mobile phones. This solution hugely reduced entry barriers to the industry and enabled downstream supply-chain innovation with their emerging ideas. Now every year almost 200 million shipments of mobile phones are based on MTK's single-chip solution, which accounts for 20 per cent of the world market. As MTK was a latecomer in the mobile and mobile computing industry, MTK was still at an early stage of building up its business ecosystem.

These three typical main cases plus seven exploratory cases aim to demonstrate a rich case base with a comprehensive view.

1)   Number of cases

A single case study relies on specific events such as testing a well-formed theory, representing a unique case, or a typical case condition, and can be a revelatory case and/or a longitudinal case (Yin 2008). However, as the aim of this research is to deliver the general nurturing process of the business ecosystem, multiple case studies should be adopted. Multiple studies can provide more compelling evidence, and therefore the overall study is more robust (Herriott & Firestone 1983). Multiple case studies lead to a replication strategy to re-demonstrate the correctness of the potential findings. Furthermore, a literal replication will be adopted which predicts similar results from 2 or 3 main cases (Yin 2008).

In each main case three projects (sub-cases) have been selected in Table 4.2 based on the seven-case selection criteria. The projects in each main case are able to demonstrate a firm's nurturing process and the interaction between different players. In each main case, different project and product managers, marketing managers and CEOs in focal firms and ecosystem partners were interviewed in order to explore the projects deeply along the business ecosystem nurturing process.

Table 4.2   Main cases' projects list

	Projects/ sub-cases	Detail
Main Case 1: ARM	1. Mobile phone project	First milestone project to trigger the nurturing process
	2. Leader partner strategy project	Enable the ecosystem variety with big support from leader partner
	3. IP classification project	IPs categorised by specific market requirement to achieve ecosystem consolidation
Main Case 2: Intel	1. PC project	Dominator in the PC industry to demonstrate the whole process of nurturing PC ecosystem
	2. Xscale project	First trial in the mobile market in order to renew the PC business ecosystem and find new increase point
	3. Atom chip project	Initiate mobile computing device based on Atom chip and encourage partners' contribution
Main Case 3: MTK	1. VCD/DVD project	Single-chip solution with low price to win the advantage when the market was very mature. DVD is the upgrading from VCD with same strategy
	2. Shanzhai mobile project	Frugal phone concept to enable the local manufacturing network to achieve quick response, cheap price and product diversity
	3. Smartphone project	Upgrading the frugal phone, co-evolved with partners in different perspective including software and hardware side

2) Data to be collected

According to the main research questions, data to be collected is listed in Table 4.3, in order to ensure that data collection was comprehensive and to help uncover potentially important constructs. The first part helps us

Table 4.3   Data collection of research questions

Data questions	Data to be collected	Data collection instruction	Other data Source
Driving force to study the business ecosystem	The market requirement; the key requirement on network cooperation and innovation	Meeting with CEO, project manager and R&D managers	Archival, website, exploratory cases
Life-cycle and nurturing process:	Idea generation and commercialisation	Product manager regarding one project	Documentation
How to nurture the business ecosystem	Partners' involvement, product evolution	Detail project with partners inside ecosystem	Documentation, website
	Integrated supply chain	The close partners and dominant design product	Documentation, website
	Industry maturity; subversive idea; vision introduction	The project with new idea to renew the industry both from central firm or others	Documentation, website, news
Construct: What are the key constructive elements of a business ecosystem?	Key Constructive elements in each phase and general constructive elements	The key factors impacting on the project process	Documentation
	The long-term orientation factors; the short-term orientation factors. The interaction mechanism; the business and governmental environment	Detail projects learning and study the process	Documentation, news, archival
Configuration pattern: typical pattern for best practice?	The main strategy and pattern; the key constructive elements to categorise configuration	Top management team to run ecosystem	Documentation

understand the 'driving force behind the study of the business ecosystems'. The second question could describe 'how to nurture the business ecosystem'. The third is 'what are the constructive elements of a business ecosystem'. The last question is about 'the configuration pattern of a business ecosystem'.

Normally, regarding the research question: first, before doing an interview, the interview question list (as Appendix Table 1) will be developed. The interview question will tackle the area on how a focal firm nurtures its business ecosystem via sequential projects. Hence they will address three levels of interview: focal firms' corporate level, focal firms' project managers' level, as well as the ecosystem partners' level who were participating in these projects. With the help of the interview question list, the key data will be collected to answer each research question after analysis. Then the right person to interview would be identified for better data collection in regard to those three levels. All these data collected from case studies either in focal firms or ecosystem partners are regarded as first-hand data. Besides, the secondary data from archival, documentation and website are also highlighted to ensure the data triangulations so as to achieve the research of construct validity (Gibbert et al. 2008). More detailed questions are shown in the Appendix Table 1.

3) Data source and collection protocol

According to this research question, there are five steps of data collection protocol as Figure 4.3 in order to ensure the research with reliability (Gibbert et al. 2008; Lin & Zhou 2011).

First: the case selection and industry review will be done in order to get a general picture of case companies. Secondly: in each case, several key projects will be identified which can demonstrate the general nurturing process of the business ecosystem. Thirdly, these projects will be mapped in terms of milestones, key decision-making, key partners and relevant interviewees. Fourthly, interviewees will be identified with regard to a series of projects from main focal companies and their partners. The question lists should be prepared according to the four research subquestions. Every question in the list aims to develop part of the answer to a research question. Finally, after the first four steps, a single case report is produced to enable interpretation of the data in context in order to tackle the research question.

4.3.2   Data analysis methods

The overview of data analysis flow consists of three steps: data mapping, data comparison and conclusion drawing/verification. Data mapping is

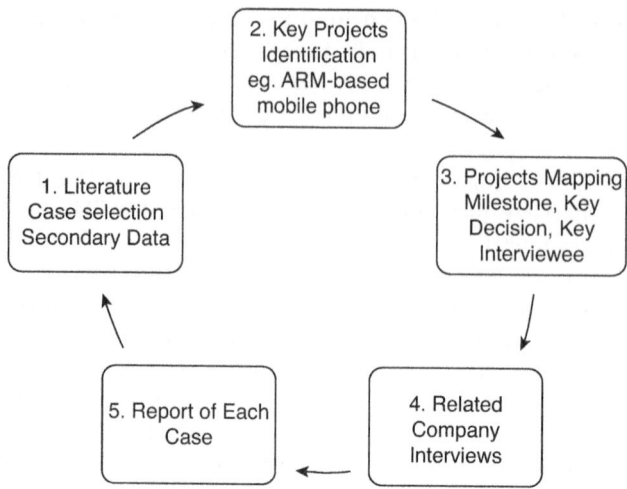

Figure 4.3 Data collection protocol

the process of simplifying, abstracting and transforming data from notes and mapping them along the time. Data comparison is the assembly of the reduced data into specific style and comparing them, from which conclusions may be drawn. Conclusion drawing/verification aims to propose the research conclusion as well as verify the result. These three steps cannot be separated but form a whole data analysis process (Miles & Huberman 1984).

We have adopted the similar data analysis methods from Miles and Huberman (1984) by linking the sequential parts of raw data (original notes), typed notes, case reports, and data analysis to research findings. In terms of data analysis method, this research enriches Miles and Huberman's (1984) three steps by the following five steps as Table 4.4: 1) to present real data by following specific patterns in terms of each research gap; 2) to conduct the intra-firm cross-cases analysis to draft preliminary findings; 3) to conduct the inter-firm cross-cases analysis to refine the draft; 4) to generalise findings by conducting cross-phases analysis (Yin 2008); 5) to build theory. Regarding the specific research gap, data analysis will just follow part of those steps. These five steps generally link the raw data with the research conclusion as well as presenting the data analysis methods. In general, those data analysis methods have a cross-case analysis nature to generalise the research findings, thus it could ensure the research with external validity (Gibbert et al. 2008).

Table 4.4 Data analysis methods

Gaps steps		Life cycle	Nurturing process	Constructive elements	Configuration pattern
Data mapping	1. Data mapping	Present the data of projects and identify the milestone activities	Present the data of projects by phase	Present the data with structure and infrastructure orientation	Present the data and select two constructive elements
Data comparison	2. Intra-firm cross-cases analysis	Preliminary findings of the draft sequential phases	Preliminary findings to draft the nurturing process in each phase	Initiate the constructs of 'structure and infrastructure' framework in each phase status	
	3. Inter-firm cross-cases analysis	Compare with other main cases analysis to refine the draft	Compare with other main cases analysis to refine the draft	Compare with other main cases to refine the initial constructive elements	Seven types of pattern determined by two selected constructive elements from cases' projects
	4. Cross-phases generalisation			Generalise the constructs cross-phases	Configuration pattern evolution
Conclusion drawing	5. Build theory	BELC by phase and phase status	Nurturing process by phases	Constructs by phase statuses and general constructs cross-phases	Seven configuration patterns and pattern evolution along the life cycle

By following the literature review and conceptual framework to collect data and analyse data, this research has internal validity (Gibbert et al. 2008). Those four research gaps in Table 4.4 would be analysed by following the enriched data analysis methods. Specifically, the following data analysis methods are explained in detail in order to draw those four research conclusions:

1) **Business ecosystem life cycle (BELC) analysis method**: this method is to identify the sequential phases of the BELC. This method has the following steps:
   - To present the project data and identify the key milestone activities.
   - To combine milestone activities of a similar nature and to highlight the initiation of phases.
   - To make comparisons between the three firms and improve the findings of phases.
   - To identify the sequential phases with their key features as well as their phase-ending status.
2) **Business ecosystem constructs analysis method**: this method will deliver the constructive elements of a business ecosystem. This method consists of the following steps:
   - To categorise the data in each life-cycle phase-ending status by using the 'structure and infrastructure' framework.
   - To compare the data of each project within each main case and identify the constructs of the business ecosystem.
   - To assemble the preliminary findings of 'structure and infrastructure' and then position the constructs by phase status.
   - To categorise the constructive elements according to their natures by phase status and then generalise each constructive element.
   - To define generalised constructive elements of the business ecosystem
3) **Business ecosystem configuration pattern analysis method**: this method will identify the typical configuration pattern of a business ecosystem. This method consists of the following steps:
   - To present data regarding to the BELC phases and then select two critical constructive elements for categorising configuration patterns.
   - To use the selected elements to measure those data by phases and deliver categorised configuration patterns.
   - To compare the adoption of those configuration patterns and deliver the general configuration pattern evolution models.
   - To draw conclusions as to what are the typical configuration patterns of the business ecosystem and its evolutional model along the BELC.

	Research Preparation			Data Synthesis		Theory Building		
	Industrial observation	Literature review	Research gap	Case studies	Data analysis	Theory building	Practical implication	Writing up
Driving Force	Identify most outstanding issues			Seven exploratory cases		Internal and external driving force		Industry challenges
Life Cycle		Some literatures	What	Three main cases with sub-cases to demonstrate process	Initiate draft phases	Life cycle by phases	When to participate in the business ecosystem	Life cycle by phases
Construct		Lack of literature	What	The key concerned dimensions	Analyse the key elements	Structure-infrastructure elements	To form a business ecosystem	Figure out the business ecosystem constructs
Configuration Pattern		Lack of literature; determined by constructs elements	What	All exploratory and main cases to identify pattern	Find the two most outstanding elements to identify configuration	Different patterns of configuration	The best practice: specific objective and operations	Typical pattern and pattern evolution model
Nurturing Process		Lack of literature	How	Three main cases with sub-cases to demonstrate process	Draw the process	Phases of nurturing process	How to participate	Nurturing process by phases

Figure 4.4 Overview of procedure

4) **Business ecosystem nurturing process analysis method:** this method will deliver the firm's nurturing activities under the context of business ecosystem life cycle phases so as to achieve the firm's strategic objectives. The method consists of the following steps:
   - To present the data in each phase of BELC.
   - To compare the nurturing process of each project within each main case and identify each firm's typical nurturing process.
   - To assemble those main cases' projects by BELC phases and compare them to deliver the generalised nurturing process.
   - To conclude what is the typical nurturing process along BELC from firm perspectives.

In conclusion, the above data analysis methods will allow the research conclusions to emerge from the data.

4.3.3   Research procedure overview

Figure 4.4 is an overview of the research procedure divided into three main research stages: research preparation, data synthesis and theory building. Coloured areas show where activities take place.

The research preparation stage includes industrial observation, literature review, research questions and gap identification. In the second main stages, case studies and data collection cover all the potential research gaps. In the last, theory-building stage, data analysis is followed by the theory-building process which addresses the research questions, and finally proposes the practical implications. Thus, all processes aim to consolidate the new theory findings and to deliver the study.

4.4   Conclusion

This chapter has explained the research questions, framework, objectives, research methodology and research design. Four aspects of the business ecosystem theory are to be studied including the life cycle, constructs, configuration pattern and nurturing process as a whole.

In-depth multiple case studies are adopted as the key methods in the qualitative research. Three main cases are selected based on seven criteria followed by the data source, data collection and protocol.

The five steps data analysis methods are developed based on Miles and Huberman's (1984) work, and this is followed by a research procedure overview.

Part II

Case Observation of Business Ecosystems

Following Chapter 4 of the research methodology, Part II introduces the main case studies and aims to demonstrate each case's business ecosystem evolution relating to those four research questions.

Each main case will follow two steps: first, three key projects in each main case are selected which are able to demonstrate their own business ecosystems' development. Secondly, the data analysis method of Table 4.4 will be used to initiate the sequential phases of the BELC.

Chapter 5   ARM
Chapter 6   Intel
Chapter 7   MTK

5
ARM Nurtures the Business Ecosystem from the Beginning

5.1 Introduction

5.1.1 Company background and products

ARM is the world's leading semiconductor IP (intellectual property) supplier. IP[1] is designed to generate a specific function, which is equivalent to the heart of semiconductor chips. ARM has offered the IP license business model to those IC design companies. Following the business model, IC design companies develop their chips by combining different IPs from ARM with their own design to deliver the chips. ARM charges license fees to partners using ARM's IP to design other chips. ARM also gets a royalty fee when partners ship ARM-IP–based chips. In order to attract more users, ARM set up a 'connected community' in order to provide their partners with a range of tools, software and systems IP to facilitate adoption and incorporation (Finlay 2002). ARM also licenses the IPs to chip manufacturers to improve their technique for manufacturing ARM-based chips.[2] ARM-IP–based chips are used by different OEMs for different digital markets including the mobile, computing, embedded and home market.

So far, ARM has already developed their new IPs from ARM7, ARM9, ARM11 and then to Cortex A, Cortex R and Cortex M as well as its technical architecture evolution as Table 5.1. Meanwhile, ARM also used these IPs to penetrate different markets from the mobile to home, embedded and enterprise markets as Table 5.2.

In the industry review chapter, the architecture/platform selection is discussed in terms of facing a new requirement in the mobile computing industry. ARM's architecture is one of the potential solutions while the other is Intel's architecture. Seen from Table 5.1, ARM's architecture was developed since V4 with a 16-bit thumb set. Later more and more

Table 5.1 ARM's product evolution

Product	Announced	Architecture	Specification
ARM7	1993	V4	Thumb
ARM9	1997	V5 (part is V4)	VFPv2; Jazelle
ARM11	2002	V6	TrustZone; Thumb2; SIMD
Cortex M	2004	V7	Thumb-2 only
Cortex A	2005	V7	Dynamic Compiler Support; VFPv3; NEON™ Adv SIMD
Cortex R	2005	V7	Dynamic Compiler Support; VFPv3; NEON™ Adv SIMD

Table 5.2 ARM's market share

Market Status	Mobile	Home	Embedded	Enterprise	Mobile computing
Market share	98%	50%	Limited	Limited	Growing
Strategy	Maintain	Enlarge	New increase point	New increase point	Just penetrate
Standard	High	Medium, against with MIPS	Highly different	Highly different	Against Intel's architecture
Competitor	Intel	MIPS	Microchip 8 bit; NEC; Renesas; Freescale; Infineon; TI	Many	Mainly Intel

functions were added into this architecture: Java processing function was added into the V5 architecture; later Thumb-2 and security functions (TrustZone) were added in the V6 architecture; Neon function was added in V7 architecture. IPs have been developing all the time based on different architectures.

Seen from Table 5.2, ARM is the dominant design in the mobile market as 98 per cent of the mobile market is using ARM IPs. The home market is ARM's second largest market which includes digital TV, media player, set-top box and so on. In the embedded and enterprise market, the market is very fragmented as many other competitors are also occupying market share. In the emerging mobile computing industry, following the competitive advantage from the mobile world, ARM's architecture is regarded as one of the best solutions for future devices. However, the other potential solution is Intel's processor, which is the dominating standard in the PC market with a strong computing capability. As a result, in the mobile computing industry the architecture solutions are focusing on these two architectures: ARM or Intel.

5.1.2　General evolutionary path

Generally, ARM started its IP license business model in 1993, and successfully commercialised its IP from a mobile phone project in 1997, then ARM continued to improve their IP performance generation by generation. ARM also realised it was very important to cooperate with partners around the product supply chain, not only with direct customers and suppliers. As a result, ARM began to set up a connected community to attract different types of partners. In 2004, ARM found that the embedded market was very specialised, and their IPs were not customised enough for these markets. So ARM began to categorise their IPs into three streams: Cortex A, Cortex R and Cortex M in order to cope with the mobile computing market, industrial control and low-level embedded market respectively.

5.1.3　Key challenges in the mobile computing industry

ARM as an IP provider is positioned upstream in the value chain and far from the end-user in the mobile computing industry. There are four challenges for ARM's business.

The first challenge is how to promote its IPs and persuade IC design firms to use them since there are many competitors in the mobile computing industry.

The second challenge is to optimise ARM's IPs' capability and performance from the software vendors' side. Usually, IC design companies cannot make full use of ARM's IPs because the bottleneck is that software is not well designed and customised for ARM's IPs. The end-user cannot experience the improvement of ARM IP's evolution. So software vendors are also key partners promoting ARM's IP, and ARM has to engage them in ARM's IP development as well.

The third challenge is how to reduce lead time and enable foundry players to easily manufacture chips. To accomplish this, ARM has to cooperate with design tool providers and foundry players. These three kinds of companies have to share their IP libraries, which would make it easier for ARM engineers to design IP and for foundry companies to manufacture ARM-IP–based chips.

The fourth challenge is to enable downstream partners' innovations. As the end-user products are still not finalised in mobile computing, collective actions among partners are encouraged to approach the best design.

In the following section, which is about the general evolutionary path of ARM's success, three projects are selected and will be deeply studied to demonstrate how ARM built up the business ecosystem regarding the four key challenges. Figure 5.1 listed those three projects. ARM7 and ARM9 projects are intended to demonstrate ARM's first success in the mobile phone project and it initiated its business ecosystem from the beginning. The 'leader partners strategy' project is to show how ARM continued to develop new IPs and maintain the business ecosystem as it grew larger. The third project – IP categorisation – demonstrates how ARM optimises their partners to meet the requirements of specialised markets.

5.2 Project 1a: mobile phone (ARM7 and ARM9)

ARM was built up through cooperation between Apple Computers and Acorn, as Apple wanted a low-power microprocessor for its handheld device – the Newton notepad. Although this device was not successful in commercialisation, it helped ARM establish its license business model. With that experience of the Newton notepad, ARM began to address the growing mobile phone market[3] for low-cost, low-power, high-perform-ance devices (Garnsey et al. 2008).

There were also many similar industrial competitors who were concerned about the next generation of mobile devices since 1993. None of them could demonstrate the future needs very well. At that time, ARM's product – ARM7 – had low power consumption and was of small size compared with other competitors' processor core. ARM aimed to be the top player in the market. Meanwhile, TI as a chip manufacturer was also trying to identify a firm that could provide a low-power proc-essor as they were not very strong in this area. However, ARM as a small company could not convince TI completely to adopt its architecture.

So ARM reconsidered its position and took a different way to persuade TI to adopt its architecture. Normally, NPD (new product development) was implemented along the supply chain from upstream to downstream. Regarded as far from the end-user customers, the key difficulty for ARM was to convince IC design firms to adopt their IPs and then create a new supply chain to produce end-user products. After being rejected by TI seen from Figure 5.2, ARM first began to connect with Nokia (a Finnish company) for help, seen as the upper arrow. At that time, Nokia also hoped to develop the next generation of mobile devices with low power

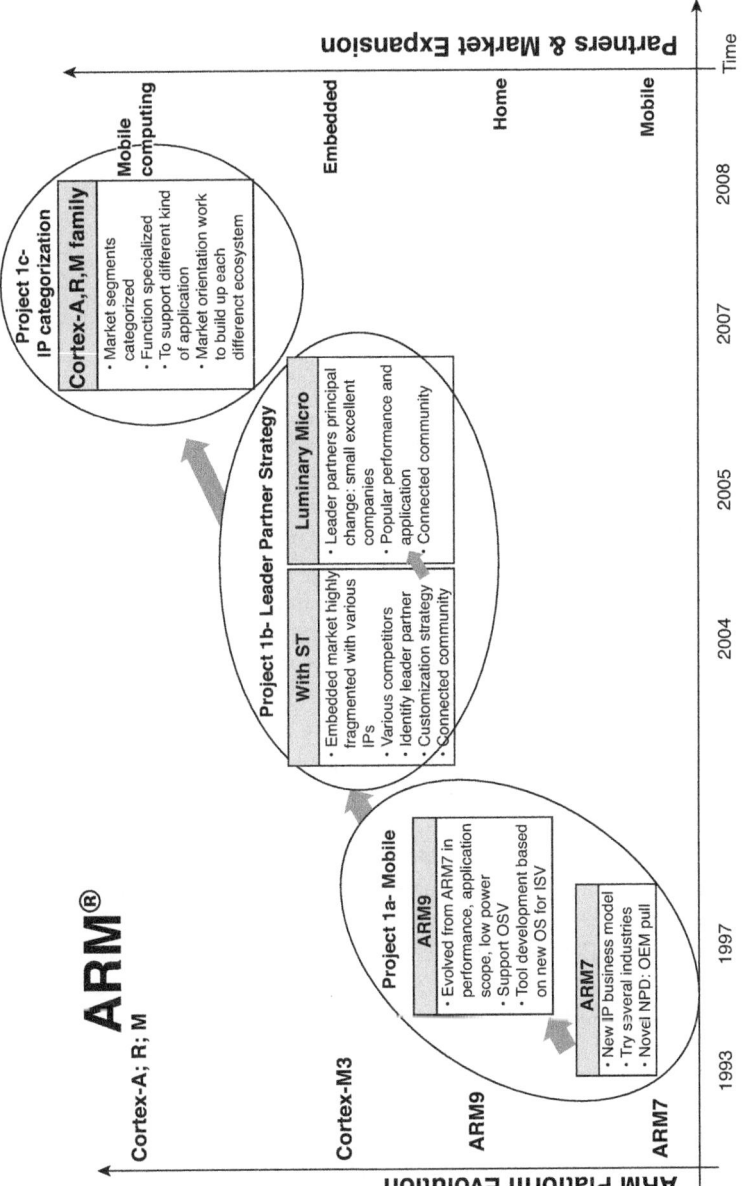

Figure 5.1 ARM's nurturing process of their business ecosystem

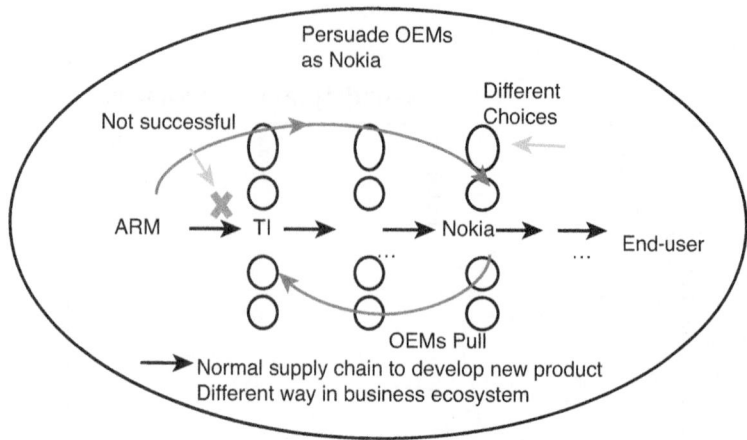

Figure 5.2   ARM's first mobile project with NOKIA and TI

consumption, small size and easy upgrading. ARM's architecture exactly matched their aims, and Nokia agreed to develop the new phone based on the ARM7 architecture.

TI was originally one of the key chip suppliers for Nokia's mobile. After Nokia cooperated with ARM, Nokia suggested TI adopt ARM's architecture for future product development. Finally these three companies built up a consortium to work together for the next generation of mobile phones. ARM succeeded in building up the new supply chain for their IPs. The first product, Nokia 6110, was announced in 1997. Then, ARM's shipments dramatically increased. This project demonstrated how ARM leveraged partners to commercialise the new product. From this project, ARM began to realise the importance of OEMs and other ecosystem non-direct partners and with their help to build up the connections. They like to call this process 'OEM pull'.

However, ARM7 was the basic core for the mobile and as time went on could no longer meet the market's performance requirements. As a result, ARM proposed ARM9 based on ARM7. To broaden its application ARM9 added the Jazelle function module to deal with the Java application. The processing frequency of ARM9 was more – or less doubled from that of ARM7. Furthermore, ARM introduced the AMBA interface which made it easy for partners' peripheral processors to connect with ARM IPs. Due to the successful story of ARM7, TI continued to be the first subscriber of the new IP. With close ties to TI and Nokia, ARM kept dramatically increasing its shipments to the mobile market. Many companies followed the way and began to subscribe for ARM's IP as well.

Besides these, ARM also started to build partner network supporting programmes, they kept updates of their products and provided design tools for non-direct customers like operating system vendors and independent software vendors. ARM would involve them with IC design companies to develop operating systems and application software to maximise the performance capability of ARM IP as well as to differentiate end-use products. After 98 per cent of the mobile phone chips were based on ARM chips, ARM7 was also recycled and customised for other low-level embedded markets in order to attract more partners besides the mobile phone partners. Thus, ARM's ecosystem would penetrate into different markets and then sustain the business ecosystem growth.

In summary, this mobile project showed ARM that: it was a convenient way to start the business by leveraging the key partners (in this project, Nokia was the key player); it was very productive to cooperate closely with IC design partners as they could help ARM directly win market share; it was also productive to support the non-direct business partners who would enhance the software side of the ARM-based chips.

5.3 Project 1b: Leader Partner Strategy (LPS) for new product development

ARM was very successful in the ARM7 project. However, how to sustain this kind of innovation left ARM with a big challenge. In Project ARM7, ARM realised that IC design partners could help ARM promote its IP directly and embed IPs into different chip products. IC firms could also use ARM's IP to differentiate their own chips. With the protection and help of leader partners, ARM was not just a single IP provider, but formed a big, close-knit community with top IC design companies to compete with other IP providers and potential competitors. From working with TI ARM had learned that it was important to work closely with IC design partners. As a result, in the following years ARM selected and persuaded top IC partners and promoted their new IPs with those top IC partners before they were listed in the market. ARM called this 'leader partner strategy' (LPS). They maintained this LPS strategy after the ARM7 project. TI was the first lead partner for ARM7, ARM9 and other IP cores in the following years. The ARM9 project already demonstrated the typical process of selecting leader partners.

Seen from Table 5.1, ARM has already launched six main categories of IPs (processor cores). For each IP, the same challenge was how to alert customers to its performance and to persuade them to adopt it. To encourage partners to contribute to new IP development, ARM cooperated each time with some as leader partners. Using the LPS, ARM could get subscribers before the new IP was listed in the market and they also got support from partners to publicise the new product.

Figure 5.3 describes the general process of LPS. Five teams participated in this complex strategy. Inside ARM the architecture team worked on the basic architecture development (instruction set architecture) and the 'design team' designed the micro-architecture (IP) based on that architecture. At the same time, the 'marketing team' with the 'design team' investigated top players in each industrial sector and identified the leader partner to contribute to ARM's new IP development. The marketing team, design team and leader partners work together to find the best way to achieve the leader partners' strategy. The marketing team in this situation would convince potential leader partners using the ARM technology road map while the design team would meet the customised requirements for potential leader partners' new IP specification. Once everything was set down, the 'leader partner strategy' would be implemented. Then, in order to keep up-to-date with the leader partners and reduce lead time, ARM's modelling team would provide a simulation model of the new IP before its release to help leader partners develop their own product in advance. The leader partners strategy has been used successfully by ARM to promote several different IPs including ARM9, ARM11, Cortex M series, and their business ecosystem has grown dramatically.

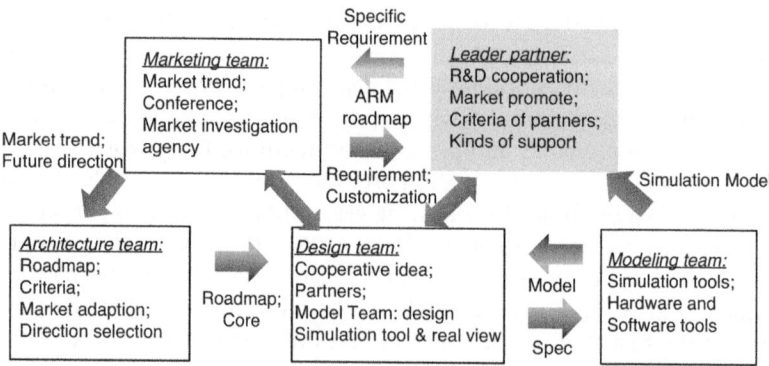

Figure 5.3   ARM's leader partner strategy for new product development

In the following section, two projects have been listed to demonstrate the success of the leader partner strategy: Cortex M3 with Luminary Micro and ST. (ARM7 was recycled and customised to be the new IP: Cortex M3.)

5.3.1 Cortex M3 with Luminary Micro

The embedded microcontroller market was huge, with over 26 billion USD annual revenue, but was also extremely splintered, with more than 40 suppliers feeding in more than 50 architectures (Booth 2007) and with many sub-sectors. ARM in 2004 initiated a new IP – Cortex M3 to meet those dynamic requirements.

Normally, ARM would select one of the top players in each sector as its lead partner, e.g. ARM9's lead partner – TI. However, sometimes they have to compromise. Since the embedded market was very fragmented, ARM's leader partner selection principles changed slightly. Instead of focusing on top players ARM had to find a company that matched well with ARM's strategy. Luminary Micro was just the case.

Luminary Micro was originally a small but active player that was to design, market and sell ARM-based microcontrollers within the popular industry market. With its huge growth and expertise in the embedded market, ARM invited it to be one of the leader partners even though it was a small company.

At that time there were three streams of MCUs (Micro Controller Units) in the market: 8bit, 16bit and 32bit. Industrial partners were hesitant about which streams to choose. However, four kinds of requirements from the market side were pushing practitioners to decide: higher performance with more applications; fast delivery times; easy connection with other systems; cost savings (Booth 2008).

In order to meet the industrial concerns, Luminary Micro with ARM developed the series processor: Stellaris family. Stellaris based on Cortex M3 offered a direct path to the strongest ecosystem of software/hardware tool development as shown in Figure 5.4.

5.3.2 Cortex M3 with ST

In order to expand Cortex M3's market share, ARM also followed their original principles to identify more leader partners. For example, in 2006, ARM also aimed to push the leading semiconductor players into the embedded market using the Cortex M3 core. Unfortunately, none of the top five companies wanted to be the lead partners with ARM for Cortex M3 development. So ARM compromised to take ST[4] as the backup. However, ST was also in a dilemma in adopting ARM's

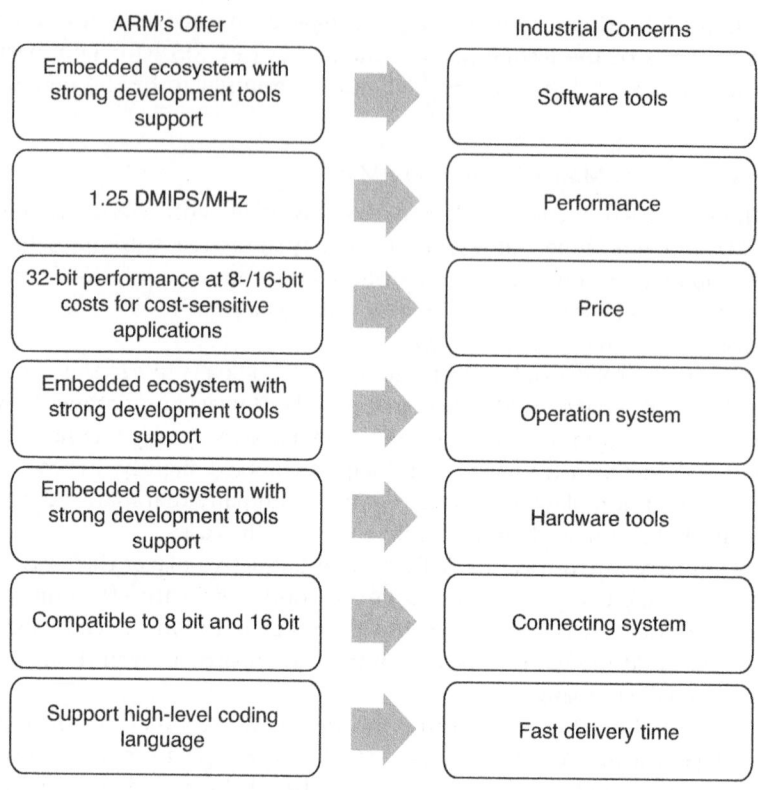

Figure 5.4 ARM's offer for industrial concerns

architecture because ST's microcontroller chips were based on 16bit data processing not 32bit data processing like ARM's. However, ST predicted that in the future, more and more functions and performance should be achieved by 32bit processors, which double the data processing capability of 16bit processors. The era of pervasive 32bit computing, control and communication would be coming soon.

As seen in Figure 5.3, after the marketing team from ARM convinced ST, the architecture and design team from ARM combined two coding standards: 16bit and 32bit together in order to let the new ST processor (ST MCU) be compatible with previous 16bit processors as in Figure 5.5. Furthermore, the modelling team from ARM also provided relevant design tools for ST to speed up their MCU development. As a result, ARM's customisation strategy for leader partners took effect and ST finally subscribed as the leader partner.

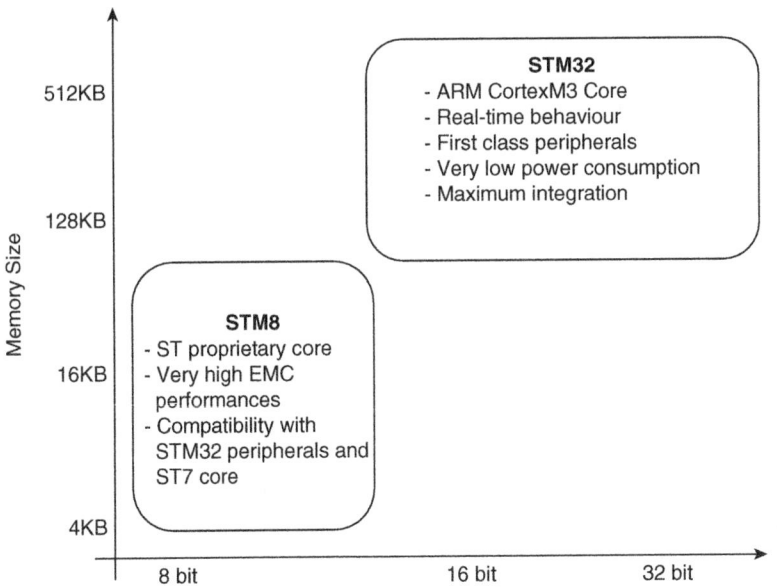

Figure 5.5   ST Micro-controller unit road map with ARM's Cortex M3

In summary, from the three projects of ARM9, Cortex M3 with Luminary Micro and ST, there are three steps to select leader partners: 1) for the marketing team to identify the top player in each market sector (TI and ST), or the players that have strong expertise in a specific area for IP development (like small company Luminary Micro); 2) for the architecture and design teams to get requirements from the leader partner and embed them into the new IP development; 3) for the modelling team to reduce the lead time of leader partner product by providing the IP simulation; 4) for the marketing team to work with leader partners to promote the new IP together by highlighting ARM's strong business ecosystem support.

5.3.3   Connected community strategy

Besides the projects of the leader partner strategy, ARM also set up a new department called 'Connected Community' to support its business ecosystem. In this community, as ARM opened its architecture to some degree or offered relevant design tools to all the partners, so every partner is encouraged to contribute their different ideas to ARM's IPs and to co-develop the end-user products. In their connected community, there are three categories of partners: silicon partners (the direct business customers), design support partners (the suppliers of design tools to ARM) and software, training and consortia partners (the partners to enhance the performance of ARM's IPs). So far, there are more than 600 ARM licensees and partners from across the ARM value chain.

The first category, 'silicon partners' consists of OEM, IC design companies and pure foundries players. ARM licenses its technology to silicon partners. These companies take ARM's IP and develop them into digital electronic products for the mass market. They have a direct business relationship.

The second category, 'design support partners', are mostly EDA tools providers, whose tools enable ARM engineers to design ARM IP architectures to high levels of performance and manufacturing feasibility.

The third category contains software, training and consortia partners. Not all of these companies do business directly with ARM, however. ARM's architecture performs better with their support from the application side and customers' side. On the software side, there are operating system vendors, application software vendors and middleware vendors. Actually all of these vendors cooperate with chip design companies directly, who are the direct customers of ARM. Training partners will help ARM to train the engineers who are working with ARM's technology and highlight their technology development. Regarding consortia partners, ARM participates in a number of industry organisations in an effort to drive innovation, create and explore new market opportunities for their partners of the connected community. With the strong support from this connected community, ARM provided not only the IP itself, but also support with design tools, operating systems and application software. These various connections will stimulate further innovation opportunities.

5.4   Project 1c: ARM IP Categorisation

ARM developed its IPs following the single road map from ARM7, ARM9, ARM10 and ARM11 in order to cope with the growing performance requirements from the leading markets. After almost 15 years' effort, ARM has already built up a huge business ecosystem. However, this huge business ecosystem does not work perfectly. For example, ARM was good at meeting the requirements of the top mobile phone market, but they did not greatly succeed with their IPs customised for other low-level markets as well because after 2000, the embedded market was divided into several segments with various functional requirements: the mobile market was the biggest market in the embedded market, while home entertainment and the facilities market was also emerging as very popular. Furthermore, industrial real-time control and automobile and medical instruments as part of the embedded market were also very demanding. In view of the industrial background, and having 15 years of industrial experience, ARM decided to categorise their new IP development in terms of different market needs based on the classic IP cores.

From 2004, ARM began to classify its cores into three groups: Cortex A series, Cortex R series and Cortex M series according to market area.[5] Cortex A series with high performance was aiming at the mobile and mobile computing industries. In this area, processors need the capability to run at very high frequency and deliver around 5000 DMIPS of performance per core. The Cortex A series also contain VFPv3 (floating point operations), Jazelle (Java application), Trustzone (security function), and NEON (accelerate multimedia). The Cortex M series are energy-efficient, easy-to-use processors designed to help design firms meet the needs of future embedded applications by delivering more features at a lower cost, increasing connectivity, better code reuse and improved energy efficiency. They have fewer function modules compared with the Cortex A series. The Cortex M series are often used in automotive, white goods, consumer products and medical instrumentation. The Cortex R series are real-time processors designed to help chip design firms meet fast response needs and are used in industrial control and other high-reliability systems. After the categorisation, ARM also followed the leader partner strategy to promote the new IPs.

In summary, the IP categorisation reflected how ARM turned its business ecosystem from a very complicated and disordered structure into a specialised and optimised one. This IP classification is matched to the different streams of specialised markets, which also helps partners in the business ecosystem to converge around those specific markets. This process finally helped divide the complicated and huge business ecosystem into many smaller business ecosystems with more specialised capability and products.

5.5   ARM's evaluation and its business ecosystem development

ARM led a very open attitude on nurturing its ecosystem since its IP was the bottom level technology of electronic devices and its position is very far from end-user products. Table 5.3 described the routine that ARM used to nurture its ecosystem. Three projects were explored to demonstrate the typical process for the growth of ARM's ecosystem. By using the data analysis method in Chapter 4, five phases could be identified that compose the BELC and explain the context in which ARM implemented those strategies.

Normally, in the first phase of a business ecosystem, there is an emerging solution for the new industry. ARM not only provided that solution but also had to leverage supply-chain partners and create a supply chain for that solution. ARM7 and ARM9 could be regarded as key solutions for the emerging industry and ARM persuaded Nokia and TI and highlighted the importance of creating the supply chain.

In the second phase, a solution from both ARM and its competitors was making the ecosystem very diversified. In that time, on the one hand, ARM launched the leader partner strategy to promote its new IPs, and on the other hand built up a connected community and encouraged partners to contribute their parts. As a result, many firms working with ARM made the industry in this phase appear full of collaboration with diversified ideas emerging.

The third phase demonstrates that after the leader partners' strategy, many leader partners were already winning competitive advantage in the market. For example, TI's chips were very popular in the mobile market, and ST's MCU chips were well accepted in the embedded market. As a result, the markets became specialised and strong alliances were also formed in order to strengthen the solutions. In this phase, partners were already positioning themselves and the solutions also converged.

In the fourth phase, after many IC design partners won market advantage, ARM also hoped to maintain its position as their key supplier. So ARM continued to improve the performance of its IPs as well as providing many supporting tools and activities. All these strategies enabled ARM to maintain its relationship with top players. ARM began to finalise its design while IC design partners began to differentiate their end-user products based on ARM's offer. As a result, ARM's business ecosystem became stronger and more robust.

The fifth phase presents the new increase triggered by ARM's new IP. In order to penetrate into other markets and sustain the ecosystem growth, ARM7 was recycled and redesigned for the low-level embedded market. Thus, other partners besides mobile phone partners would join ARM's business ecosystem and then trigger other new product developments. As a result, ARM's business ecosystem would experience growth again along the BELC.

In a later section, the data analysis method of Table 4.4 will be continued. The projects from Intel and MTK will add more content to these draft five phases.

Table 5.3 ARM's business ecosystem development by phase

Projects	Business ecosystem evolution			
1a – Mobile phone	1. ARM7 as new solution for industry: low power IP 2. ARM persuaded Nokia and building up consortium with TI to leverage supply chain partners	1. Solution Co-develop (ARM7 and ARM9 with TI) 2. IC design firms could differentiate their products by using ARM's IPs as they were very open. 3.Expand relationship with 3rd software supplier to help enhance ARM's IP	1. Strong alliance with Nokia and TI, other IC design firms followed 2. TI's chips winning the competitive advantage	1. Continue to improve ARM7 and ARM9's performance 2. Keep supporting IC design partners	1. ARM7 was customised for the embedded market 2. New IP to penetrate into other markets
1b – Leader partner strategy		1. The huge embedded market & IP architecture diversity 2. Partners collaboration: leader partners strategy: co-design, co-marketing 3. Third party support the key platform: connected community	1. More than 1000 customer-chosen STM32 (Cortex-based); 2. Strong software ecosystem	1. More partners joining as the leader partners	

Continued

Table 5.3 Continued

Projects	Business ecosystem evolution				
1c – IP categorisation			1. The embedded market specialisation 2. ARM's IPs customised for markets 3. ARM's business ecosystem complicated	1. Three streams of IP identification 2. ARM's business ecosystem divided by these three streams	
Phase ending status from ARM	1. New solution for emerging industry 2. Simple supply chain initiated	1. ARM IP made industry solution diversity 2. Industry highly collaboration and flexible connection	1. Market specialisation and convergence 2. Supply chain finalised	1. Solution continue improving 2. Supporting fixed relationship	1. Solution upgraded
Phases	Phase 1	Phase 2	Phase 3	Phase 4	Phase 5

6
Intel Re-Enters the Mobile Computing Business Ecosystem

6.1 Introduction

6.1.1 Company background and product

Intel is an IDM (Integrated Device Manufacturer) company and is the top player in the semiconductor industry. Intel was founded on July 18, 1968, based in Santa Clara, California. Intel mostly focused on the computer processor unit (CPU) and also made motherboard chipsets, network interface controllers, flash memory, graphic chips, embedded processors and so on.[1] In 2010, Intel had around 70 per cent of the world processor market, with the 2nd largest manufacturer, AMD, having around 26 per cent. The rest only accounted for very little market share.

6.1.2 General evolutionary path

Intel has successfully dominated the PC industry for more than 20 years since the 1990s (Gawer & Cusumano 2002). In the early 2000s, they began to penetrate into the mobile computing industry which was a huge market. So far there are two main computing architectures: RISC led by ARM focuses on embedded processors while CISC led by Intel (such as X86) focuses on the computer industry. But now both of these groups' companies are moving into the mobile computing industry which is arising from a convergence trend between the mobile and computing industries.

Intel realised that the growth would be triggered by the consumer electronics market. Consumer electronics will have perhaps ten times as many shipments as the PC industry. However, Intel was not a strong player in this industry even though they had an ARM-based IP (Xscale processor) through acquiring other firms. They had tried to penetrate the market in 2001 with IP, but unfortunately Intel failed and withdrew from the mobile industry in 2006.

Then Intel changed its direct approach into the mobile market and proposed a new strategy to enter the mobile computing industry with its own architecture processor – Atom – in 2008. In order to highlight the Atom processor for mobile computing devices, Intel also set up the Moblin ecosystem for operating system development to enhance Atom's performance. In all, Intel planned to reach the mobile market step by step along with the establishment of their business ecosystem.

6.1.3   Key challenges in the mobile computing industry

Even while Intel was dominant in the PC industry, they faced three big challenges when they started to penetrate the mobile world. Their first challenge was about technology performance: so far Atom chips did not reach the level of low power consumption and tiny size. The second challenge was how to gain sufficient partner support for Atom chips to challenge the big competitor, ARM. There are still not enough partners to use Atom chips. Normally, an OEM will select chips by considering not only the chips' capability themselves, but also the degree of support from software partners. The third challenge was about manufacturing service. In order to shorten lead time, partners hope to get chips not only from Intel's foundry plant but also from other foundry players. However, so far no other foundry player could wholly manufacture Intel's chips.

In the next section, a historical overview of Intel will be conducted in order to describe how Intel nurtured its development into the PC and mobile computing industry as shown in Figure 6.1. There are three of the most typical projects to demonstrate the process. First, the successful PC projects will be reviewed to address the whole BELC. Secondly, Intel launched a project named Xscale based on ARM architecture in order to enter the mobile phone market. This project was not successful, from which many points were learned. Thirdly, Intel changed the direct way into the mobile market by using the Atom processor with comprehensive support from the software side (Moblin ecosystem) and market side (initiating new application concepts like Netbook and MID).

6.2   Project 2a: the PC industry

Since the 1990s, Intel dominated the PC market through their strong and robust business ecosystem around the generation of processors. Intel implemented a sequential four steps/principles to build up its business ecosystem successfully.

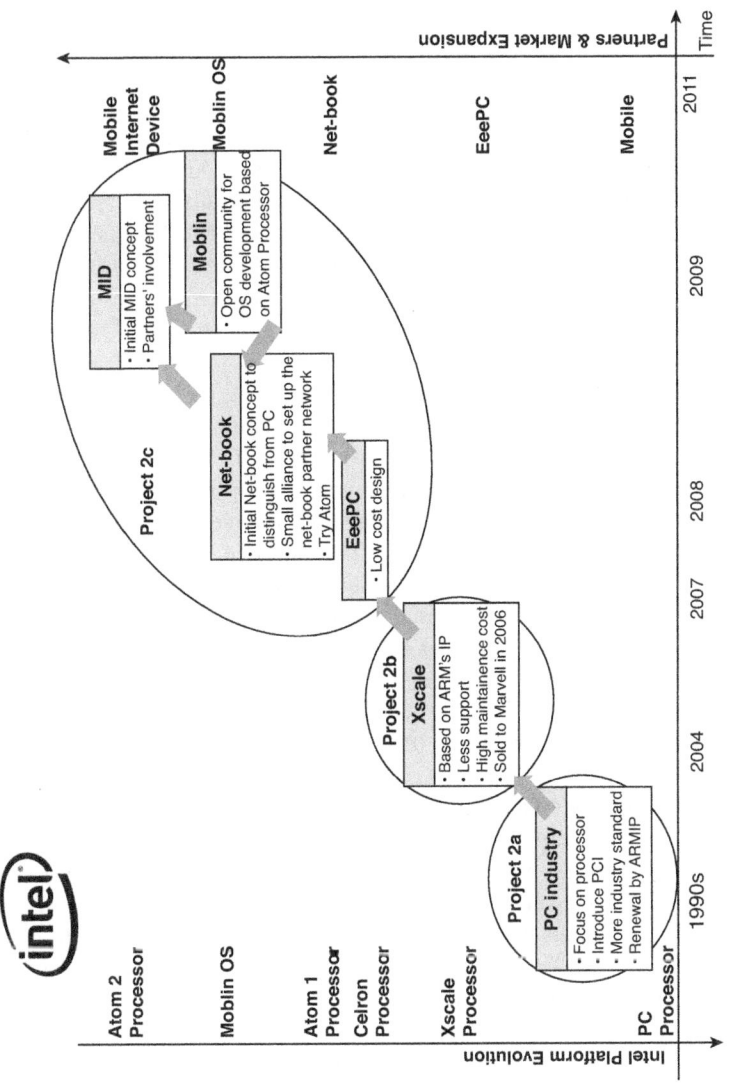

Figure 6.1 Intel's nurturing process of their business ecosystem

6.2.1   Be a key supplier to IBM

In the early 1980s, Intel had two main types of business: memory chips and processors. However, for two reasons Intel positioned fully in processor development: in the early 1980s, Japanese firms won competitive advantage with excellent business models; IBM was very successful in the PC industry and Intel was one of its suppliers.

At that time IBM and Apple demonstrated two different PC industry business models: Apple was very vertically integrated while IBM extended Apple's model to build a community of suppliers. Due to the openness of IBM, many niche players like Intel and Microsoft fully supported IBM's PC development. In 1986, IBM dominated the PC market with 5.65 billion dollars in market sales, which was also followed by Compaq, a big player with an even more open attitude (Moore 1993).

6.2.2   Platform leadership strategy

IBM dominated the PC industry by introducing its own type of PC product design. This standard design set strict rules on hardware, software and other peripherals. Every component supplier had to follow its specifications. IBM also charged partners for adopting its PC design. In this way IBM set high entry barriers for small players, and also reduced the product innovation capability of key suppliers. For example, IBM's PC design in the later 1980s was already an obstacle and could not make full use of Intel's processors. To make things worse, there were also many other types of PC design besides IBM's which industrial suppliers spent time and investment in adapting to. Moreover, all of these designs were charged with an IP fee.

Intel wanted to remove these IP fees and allow suppliers to work more flexibly and efficiently, which would also simplify partners' connection with its processors. In the early 1990s Intel began to set up an industrial standard – PCI (Peripheral Component Interconnect) – for free, which was an interface for its processors to connect peripheral components from other partners (Gawer & Cusumano 2002). Actually, other peripheral component suppliers were also facing a compatibility problem with IBM designs. Intel's industry standard let them freely connect to other partners. By using Intel's standard, industrial partners could get rid of IBM's restrictions and controls to assemble PCs their own way. Furthermore, Intel also promised that PCI would continue to make peripheral processors or devices compatible even when the processors were upgraded. With the introduction of PCI, the PC industry began to specialise itself and ceased to be dominated by OEMs (like IBM and Compaq) being instead coordinated by key platform providers like Intel.

Besides this, Intel was still continuing to set up other promising indus-trial standards: USB (Universal Serial Bus) for plug-in and output devices, Wimax for next generation communication and so on.

6.2.3   The business ecosystem establishment

Intel not only set up the industrial standards, but also encouraged downstream innovation with peripheral products based on Intel's proc-essors. They supported many types of companies besides the OEMs like OSV, ISV, content providers, retailers and system integrators. This strategy would maintain Intel's big business environment and enable its growth in shipments, exceeding other leading processor providers: 1) Intel shared their processor road map with their partners and encour-aged their co-evolution to improve processors' performance; 2) Intel sent out engineers and marketing people to aid industrial partners' involvement; 3) with a strong capital background, Intel also had an investment department called Intel Capital which would provide finan-cial support to small companies or acquire companies' shares in order to stimulate their innovation; 4) besides these, the most important one was that Intel set up a close cooperation with Microsoft to finalise the dominant design (Intel's processor with Microsoft's operating system, Windows) and maintain a high entry barrier against competitors in the PC industry.

6.2.4   Technology evolution

Intel never stopped pushing its technology ahead of market needs following the pace of Moore's law: '*The number of transistors incorporated in a chip will approximately double every 24 months*'.[2] Currently Intel has already reached 28nm manufacturing technique for its semiconductor wafer process. In the current CPU or embedded processor market, 45nm or 65nm technique is already well advanced. So Intel's technology had already massively exceeded the current market requirements.

In summary, Intel has dominated the PC industry since the 1990s, because of three main reasons: 1) Intel specialised in the PC industry and focused on processor manufacturing, which was very different from IBM with a vertically integrated structure. Intel's specialisation helped it to precisely follow Moore's law and act as the technology leader; 2) besides owning the core processors, Intel also cultivated its business ecosystem, which also very much supported the commerciali-sation of the processors; 3) in the later stages, Intel allied itself closely with Microsoft to deliver the dominant design as 'Win-tel' (Windows and Intel).

By learning from its experience in the PC industry, Intel shaped its business strategy in other projects. Intel focused more on standardisation, dominant design and technology advantages.

6.3   Project 2b: Xscale

DEC,[3] as the key competitor of Intel from the early 1990s, agreed to sell StrongARM to Intel as part of a lawsuit settlement in 1997. So, with this ARM architecture, Intel penetrated the huge embedded market, especially into the mobile phone market. Intel hoped to dominate both the computing and embedded market with these two architectures: X86 and RISC. This strategy was also supported by Intel's strong business ecosystem in the PC industry.

Intel developed the Xscale project based on the ARM architecture from 2002 and was successfully winning the market competitive advantage in competing with Motorola. Then Motorola abandoned its chips and adopted Intel's Xscale chips to develop the smartphone Moto A1200. Intel also outsourced manufacturing to TSMC as they had rich experience of manufacturing ARM-based chips. This project helped Intel get considerable feedback from different perspectives including market, technology, as well as ecosystem building.

However, Intel finally sold the Xscale line of processors to Marvell in 2006 as Intel did not got significant benefit return from this project or much support for these chips. They wanted to focus on the mainstream products, so they sold the Xscale project. The reasons could be concluded in detail as follows: 1) in general, AMD as a new X86 processor manufacturer was already occupying part of the market share of Intel's mainstream products as server processors. Intel wished to maintain its dominant position in the PC industry; 2) in regard to the Xscale project, there was not as significant a profit return as its X86 processors. Also there was not much partner support on the operating system side and no mature manufacturing skill for ARM chips within the Intel plant; 3) the most important reason could be that ARM was already growing due to Intel and other big chip players' support and licenses. Especially in the mobile market, ARM architecture was in a dominant position. Intel itself had already incubated a future competitor (ARM) in the embedded market. Due to these concerns, Intel decided to abandon the Xscale project, sell it to Marvell, and dedicate itself to its own architecture for the embedded market.

This failed project also reflects some strategic bias from Intel's perspective: being hugely successful in the PC industry by closely collaborating with Microsoft, Intel thought they could reach a dominant design by copying the 'Win-tel' model. So they exactly copied this strategy into the mobile phone market. However, Intel only addressed the collaboration with Motorola, while disregarding the nurturing of the business ecosystem around the mobile phone market. Furthermore, the Mobile phone market was more complicated and fragmented, which was very different from the PC industry, in which many big players already existed and various operating system vendors as well. As a result, Intel learned an essential point: the business ecosystem should be regarded as being as important as the key partners. Then Intel adjusted its strategy for future re-entry into the mobile market.

6.4 Project 2c: Atom-based

Intel discovered a big market trend for portable devices with low power and easy Internet access. In order to cope with the booming market demand, Intel started to engage in its low-power chip development. 'Atom' was the first such chip launched in early 2008.[4] Due to Intel's technology, Atom was good at computing capability and they kept refining its low power consumption and decreasing the chip's size as well. Intel was waiting for a chance to re-enter the mobile market.

6.4.1 Asus EeePC, a new and low power solution to renew the PC industry

Originally EeePC was proposed by Asus, not Intel, but it gave Intel a good bridge to enter the mobile computing industry in a different way compared with the Xscale project. EeePC renewed the mature PC industry into a low-cost PC industry.

The Asus Eee PC was first proposed as a low-price notebook in late 2007 to demonstrate the new idea of notebook as 'easy to learn, easy to play, easy to work'. In 2008, as the global crisis highlighted views of sustainable development with low cost, people preferred a notebook with basic functions for daily use at a low price instead of full functions, possibly unuseful, at a full price. It was a big niche market to exploit.

Then, Asus began to design the potential and promising device according to the above requirements from marketing. On the technological side, Asus discovered Intel had a big stock (around 2 million units) of Celeron processor, a low-level version of processor. In order to cut the BOM cost, Asus persuaded Intel to offer a very low price to clear out its stock. After three months, Asus launched its first product – Eee PC 701 – and finally reached the 'most wanted holiday gift' in Amazon at the end of 2007. It successfully entered the North American market against huge players Dell and HP. Later, Asus also got excellent market performance in other areas. At the end of 2007, it achieved sales up to 200,000 units and still grew in the later years.[5]

6.4.2   Netbook: a new vision for low-cost PC

EeePC arose by chance, but stimulated Intel's ideas about the strategy needed to penetrate into the mobile computing industry.

Asus's success with their low-price notebook created two problems for Intel: 1) more and more other OEMs forced Intel to offer processors similar to those used by Asus; 2) the low-price notebook had a big and serious impact on normal notebook sales, and consequently cut down the whole PC industry's profit; 3) the economic and cheap products were popular in the global financial crisis. In 2008, Intel launched the next generation processor: Atom, with low power, which just matched the market requirement. In order to maintain profits and nurture the new market to the mobile computing market, Intel started to initiate a 'story' to let Atom take the place of the Celeron chip in low-price notebooks.

First, Intel abandoned the concept of low-price notebooks but initiated a new concept of 'Netbook' to demonstrate a niche market for low level of use, and to provide the new chip Atom for Netbook use. Netbooks were simple and affordable ultra-portable Internet companion devices that offered access to the Internet easily. Netbooks were aimed at first-time buyers in emerging markets and as secondary devices in established markets. The Netbook usage mostly focused on the VIEW part (like watching DVD and email) – entertainment and simple social media functions rather than heavy computing work – while notebook was focused on the DO part (like create word docs and edit photos) to achieve computing work.[6] With this story, Intel successfully categorised Netbook as a different device from notebook.

Secondly, in order to maintain Netbook's development, Intel pushed to develop the detail specification of Netbook with their OEMs partners as Table 6.1. A Netbook was for basic computing with easy Internet access while a notebook was full of computing functions. Notebook's size was more than 10.2 inches, compared with 7 to 10.2 inches for a Netbook. Netbook had less memory and a different choice of low-cost operating system. In terms of processor, Intel also highlighted that only the Atom chip would serve the Netbook market. Intel not only provided the Atom processor, but also integrated some peripheral chipsets in order to deliver a Netbook platform for OEM use.

OEMs who made Netbooks only within these features would get support from Intel with a big price advantage. If any partners broke the rules, Intel would reject the order or sell them the chips at the normal price. Intel led all of these strategies in order to set up the new Netbook market demarcated from current notebooks. Some partners initially tried to use a Linux operating system but without success and as a result continued to use Windows series operating systems.

Netbook was similar to a notebook in many ways except for the hardware and size. Intel was easily more successful than with its previous business ecosystem. As a result, Intel rapidly dominated the Netbook market in 2008 with 14.6 million units and still kept growing.[7]

6.4.3 MID and Moblin ecosystem: make the mobile computing industry sustainable

After Netbook, Intel moved further into the mobile computing industry with a new product – MID (Mobile Internet Device) described in Table 6.1. First, Intel selected several top OEMs in China based on their market performance, innovation capability and future development strategy. Then Intel selected two companies: Lenovo and Aigo (an OEM whose products varied from memory sticks to portable devices).[8] Secondly, with Aigo Intel began to develop the new concept for the mobile computing industry. Intel realised that they were not able to compete in the mobile market as they did not have that kind of processor. However, regarding the niche market (the mobile computing industry), Intel decided to initiate a brand-new market to bridge the gap between the PC and mobile industry, which would prepare Intel for the future competition against ARM's ecosystem. They brainstormed and finally developed the idea of 'MID', short for 'middle' and reflecting the gap between the PC and mobile industry. In addition, they also promulgated a new story about MID in the same way as they did with the Netbook story.

As seen in Table 6.1, MID was a portable pocket device for entertainment and information access with touchscreen or keyboard. It was to some extent more like a mobile phone than a PC. There were some challenging issues: processor; operating system, application software support and 3G license.

The MID processor aimed at all-day use. Previously Intel's processors, having high power consumption, had not been able to meet the key requirement of long battery working time (ARM had achieved all-day use in the mobile market). Intel had already introduced the new processor – first generation Atom (Menlow platform) – but it did not have

Table 6.1   Defining the different devices in the mobile computing industry

	Notebook	Netbook	MID
General view	Full mobile computing	A simple, affordable companion device for easy access to web	Information in your pocket
Usage	Productivity or homework; editing photos/video; gaming and entertainment	Internet-central applications; e-mail and note-taking; media viewing	Entertainment; staying in touch; information access
Size	>10.2 inch screen	7–10.2 inch screen	4–7 inch screen
OS	Windows	Linux or Basic Windows	Linux or Windows
Processor	Intel Centrino, Core, Celeron, Pentium	Intel Atom	Intel Atom
Weight	>1kg	<1kg	<1kg
Price	>400 dollars	250–399 dollars	700 dollars
Memory	>1G	512Mb–1G	512Mb

sufficient power consumption. However, Intel had already proposed the road map of the Atom chip for the next two generations. First Atom would aim mostly at the Netbook market and also be trialled in the MID market, while from 2009 second generation Atom (Moorestown platform) would be integrated into a two-chip platform, saving a lot of power with the target segment of MID. Later, from 2011, Intel would develop the third generation Atom (Medfield platform), which was a single-chip solution with very low power consumption. The Medfield platform would be used in the smartphone market.

Looking back at the Xscale project, one important reason for failure was that Intel did not win support from most big software suppliers. So in re-entering this market, Intel paid more attention to software enablement. In 2007, Intel set up the operating system (OS) ecosystem Moblin by adopting part of Linux open source. Moblin was a community distribution and technology development upstream open-source project which was the OS optimised for Intel Atom processor–based devices. In 2008 Intel developed Moblin 2 with advanced power management, fast cold boot, advanced UI (user interface) support and Moblin SDK (software development kit). By 2011, there were more than 15 OSV and 100 ISV in the Moblin ecosystem, and the number was still growing. Besides the open-source projects, the Moblin partners would also get financial support from Intel.

Intel intended to implement the 'Moblin ecosystem' as well as the Atom platform to fit future devices including MIDs, Netbooks, auto and other potential devices.[9] In setting up this ecosystem, Intel was more open, and co-evolved with partners both at Atom chips and software development in the Aigo project. In 2009, Intel and Aigo developed the first product, Aigo MID P888, as shown in Figure 6.2. First, they did market research with a third party agency, In-Stat,[10] to identify the top ten popular software applications for daily use. QQ was on the list. Then with the Moblin ecosystem, Intel fast developed Atom with different kinds of application such as the most popular online chat tool in China: QQ from Tecent. Aigo, as an OEM, also promoted its product with Tecent. Intel also supported Tecent in adapting the new QQ version to MID. Compal, as an ODM, was in charge of designing the whole solution and final test.

There were some problems for this product: 1) the 3G network was not ready in China in 2009, which hugely cut down one advantage of this portable device; 2) the operating system was not compatible with many applications; 3) BOM cost: Intel's Atom processor was too expensive and too high power consumption. Due to these reasons, Aigo MID did not sell well (Lenovo, the other partner, quit the project because of similar reasons). Intel tried very hard to solve the second and third problems with the Moblin ecosystem and Atom improvement. They suggested Aigo adopt the operating system from the Moblin ecosystem, for which Tecent was also persuaded to develop a compatible version of QQ. The first problem (lack of adequate 3G cover) shows how the total business

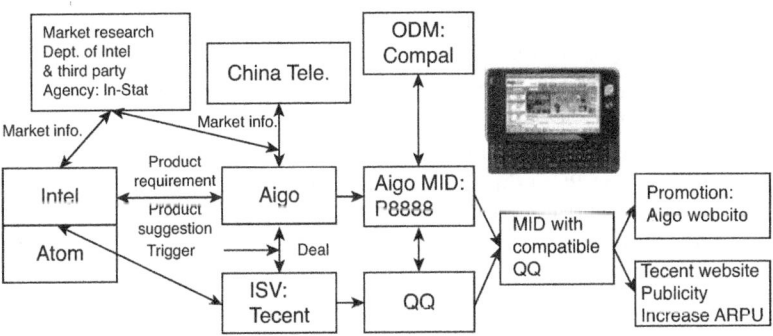

Figure 6.2 Intel's business ecosystem product: Aigo MID P888

Source: Author interview from Intel, Aigo and Tecent (China, 2009).

environment is one of the key success factors for a specific product. Intel was co-evolving with its ecosystem partners, and they hoped to develop a future solution for this brand-new industry.

In summary, the Atom-based project demonstrates how Intel tried to build up the business ecosystem for Atom chips, especially from the software side. Intel successfully attracted some partners to work with. They tried to make diversified products based on Atom chips.

6.5   Intel's evaluation and its business ecosystem development

These three projects are shown as Table 6.2 and conclusions can be drawn in comparison with the five phases of the ARM cases.

In the first phase of the life cycle, Intel did similar things as ARM: they introduced new solutions like Netbook and MID into mobile computing. They also persuaded the previous supply-chain partners to support their new strategy.

In the second phase, the PC industry was experiencing competition among different industrial architectures from IBM, Compaq and other PC OEMs. As a result, the types of computer varied. Due to industrial specialisation by key component players like Intel and Microsoft, small players were also welcome to contribute products. These processes were repeated in the mobile computing industry. Intel also proposed the Moblin ecosystem to support Atom with various operating systems and application software. However, Intel did not experience this in the second phase as they selected Motorola, aiming for a dominant design.

In the third phase, Intel introduced the PCI interface to connect different component players to achieve the dominant design for the PC industry. The standard of Netbook was also finalised by Intel.

In the fourth phase, Intel tried to maintain their advantage and continued to introduce different industrial standards to form stable structures in the PC industry. The most important thing was that Intel closely cooperated with Microsoft and delivered the dominant design as 'Win-tel' to maintain a high entry barrier for competitors and latecomers. Intel also improved manufacturing capability to enhance the performance of new processors. This phase could last long. For example, Intel has dominated the PC industry since the early 1990s.

In the fifth phase, key players would consider bringing novel products to renew the business ecosystem. At that moment, Intel persuaded their PC partners to penetrate the mobile phone market when Intel got

Table 6.2 Intel's business ecosystem development by phase

Projects	Business ecosystem evolution				
2a–PC project	1. Specialise semiconductor industry rather than Apple integration model	1. IBM model enabled the PC industry with various partners 2. Intel provide processors to support different PC OEMs	1. OEM began to build up standards, without success 2. Intel introduce PCI to connect other partners	1. Continue to introduce standard to maintain the dominating position 2. Close collaboration as 'Win-tel'	1. Develop StrongARM chip to enter the mobile market
2b–Xscale project	1. Penetrate into mobile by using Xscale project by persuading partners		1. The mobile industry crucial competition 2. Intel collaborate with Motorola	1. Many IC firms adopted ARM IP for mobile 2. Intel outsource manufacturing to TSMC	2. Finally Intel totally sold to Marvell
2c–Atom project	1. Initiate Netbook learning from EeePC 2. Introduce Atom 3. Initiate MID	1. Partners involvement and enable product diversity 2. Moblin ecosystem	1. Finalised the Netbook standards with partners		
Phase ending status from Intel	1. Introduce new solution for emerging market 2. Simple supply chain	1. Partly open hardware 2. Software partner involvement around Atom-based project	1. Market specialisation and solution convergence 2. Partner network converged	1. Dominant design 2. Fixed partners 3. Stable partner network structure	1. New/ niche market 2. Project failure and sold out
Phases	**Phase 1**	**Phase 2**	**Phase 3**	**Phase 4**	**Phase 5**

access to the ARM IP. The key idea was to renew the business ecosystem not only from a product perspective, but also from an industry structure perspective, because more partners would be involved in the mobile phone market, which would result in the specialisation of the PC industry structure.

In a later section, data analysis method of Table 4.4 will be continued. The projects from MTK will add more content to these five phases.

7
MTK Enhances the Business Ecosystem Efficiency

7.1 Introduction

7.1.1 Company background and product

MediaTek (MTK), founded in 1997,[1] was a leading IC design company for wireless communications and digital multimedia solutions. The company was a market leader and well known for its single-chip solutions for mobile chips, digital TV, DVD and VCD products. MTK got the reputation mostly because of their Turnkey solutions (single-chip solution) for mobile phone in Shenzhen region of China in 2008. MTK started this business in 2006, and became one of the biggest contributors with 20 per cent of world mobile chip shipments in south China, followed by the top players such as Qualcomm, Infineon, TI (Zhu & Shi 2010). The Turnkey solution integrated most chips inside the mobile phone board, which was presented to customers in a ready-to-use condition. The single-chip style lowered the technology barrier for SMEs with little R&D capability. Meanwhile many customers shortened the design-cycle time and reduced the design cost thereby raising their competitiveness in the market.[2] Actually, the Turnkey model was not new at all. MTK started the Turnkey model with the first of their products – the VCD chip, which was followed by DVD and mobile phone chips. The reason why Turnkey was highlighted in the mobile phone market was the huge shipment of mobile phones compared with those of the VCD or DVD markets because the mobile phones were bought by almost every individual, whereas VCD players were bought by families.

Due to lack of the design capability, not many firms could design the chips on their own. As a result, the semiconductor value chain in China was more complicated and specialised than that in the West. In Western countries the semiconductor value chain usually contained the IP provider, IC design company, foundry and OEMs which lined up clearly. Much of the R&D task had been done by single companies in the Western value chain. Figure 7.1 shows the IDHs (independent design house) which were normally small companies doing product design on the chip's platform from IC design companies. In this case, MTK, as an IC design company, tried to help and nurture the IDHs. Then many component providers or distributors would deliver these design parts to the ODM to integrate a total design for specific products. The parts manufacturing was sometimes outsourced to the EMS (electronics manufacturing service). The OEM coordinated these fragmented supply networks.

7.1.2   General evolutionary path

MTK started to penetrate the consumer electronics industry from 1997. They first integrated all the necessary chips into a single chip to cut down entry barriers to the VCD market. The VCD market was very mature as all the supply chain partners were fixed. MTK's single solution attracted many latecomers and existing players. The single-chip solution hugely cut down price and shortened product lead time which could then make unaffordable products into affordable products. Then MTK used the same strategies to enter the DVD market. Later, in 2004, MTK started to penetrate the mobile phone market. As it was a very complicated product, MTK not only provided a similar single-chip solution, but also triggered the manufacturing network in the Shenzhen area. This network scaled up the MTK's chip shipments and helped to rapidly deliver cheap mobile phones. In 2006, as the smartphone became popular, MTK also intended to expand itself into this emerging market. However, this industry involved many more partners, and the products were more customised and complicated. As a result, MTK was faced by a big challenge.

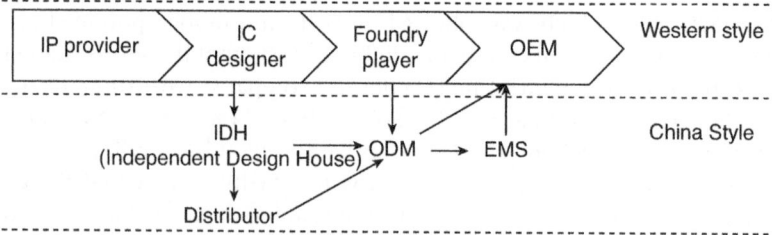

Figure 7.1   China's unique semiconductor value chain

7.1.3 Key challenges in the mobile computing industry

As a latecomer to the consumer electronic market, MTK had to catch up with existing big players in the market. Three challenges should be highlighted:

First challenge – When to penetrate a market: due to the lack of R&D capability and financial support, MTK penetrated into specific market when it was mature with the supply chain fully established.

Second challenge – How to penetrate a market: in order to win competitive advantage from existing competitors. MTK would provide the integrated solution with low cost and low entry barrier.

Third challenge – Co-evolution with partners: as the mobile phone industry entered the 3G era, more partners were involved in the mobile supply network. MTK began to co-evolve with local partners instead of only providing integrated solutions.

In the next section, a historical view will be conducted to describe how MTK nurtured its business ecosystem as Figure 7.2. There are three big projects to demonstrate MTK's general evolution. First, the successful VCD and DVD project will be reviewed to demonstrate MTK's basic strategy of single-chip solutions. Second, besides the single-chip solution strategy, MTK launched a mobile phone project to trigger downstream innovation in the mobile phone supply chain. Third, in order to meet the smartphone requirement, MTK began to co-evolve with local partners.

7.2 Project 3a: VCD and DVD market

MTK was previously a multimedia chip design department of UMC (an IDM semiconductor company in Taiwan) and was spun off as an independent IC design company in 1997. So at an early stage MTK had the capability of multimedia chip design. MTK decided to penetrate the VCD chip market with its relevant background. At that time, this market was dominated by Western companies (like OAK, Philips) and Japanese companies (like Sony). However, what they believed was the S-curve as shown in Figure 7.3. The horizontal dimension is time, the vertical is market size. The maturity of products follows the S-curve. The Western and Japanese companies already had the advantage in the early R&D stages, however the market was not yet as big as it would be in the maturity stage. MTK was waiting for that stage because they believed they were good at controlling the cost of solutions for this market. Besides, they could provide more integrated solutions to cut down industry entry barriers, as the mainstream chipset was not very integrated at that time.

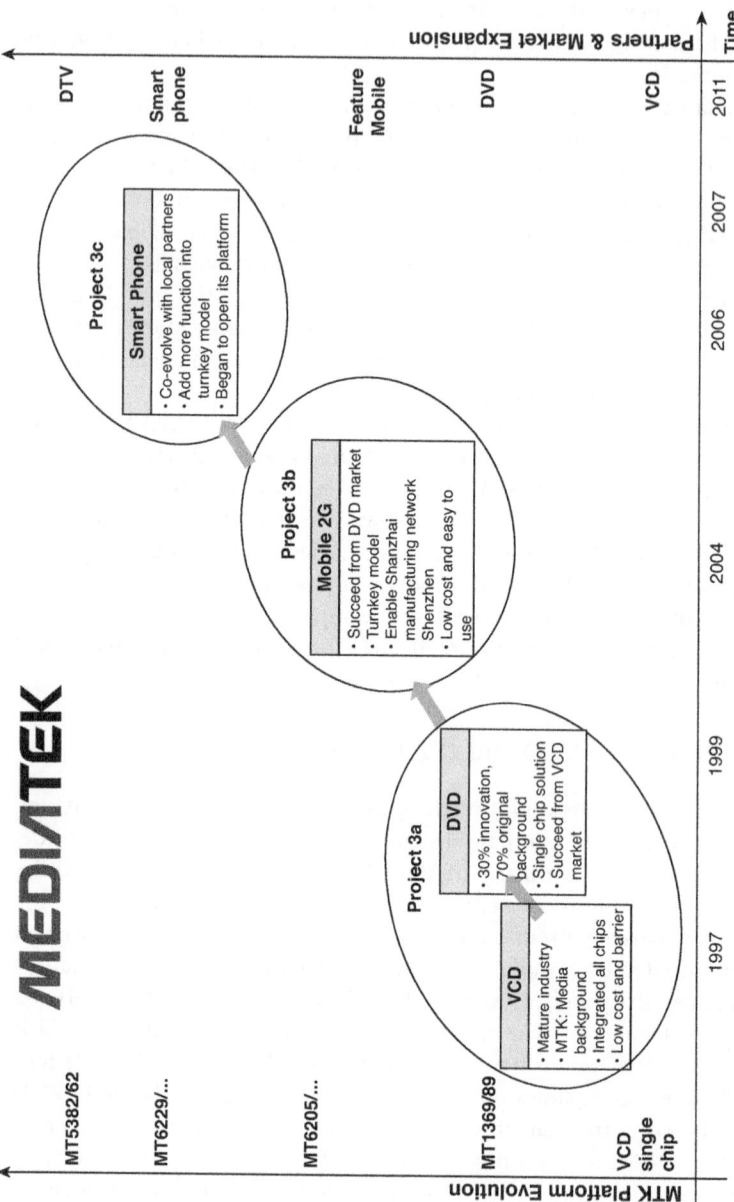

Figure 7.2 MTK's nurturing process of their business ecosystem

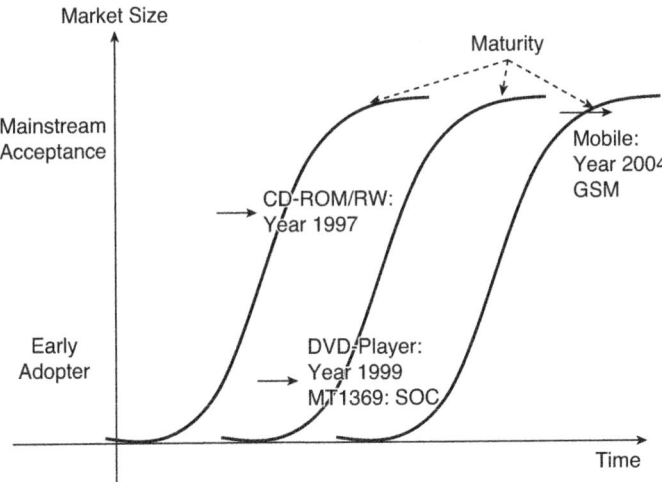

Figure 7.3   MTK's start point in different markets

MTK started research on single-chip solutions, and finally entered the market with their Turnkey solution in 1997. At that time, many factors pushed MTK to dominate this market: 1) the Turnkey model cut down the technology barrier, and enabled downstream supply-chain innovation especially in mainland China; 2) Taiwan companies formed industrial clusters in Hsinchu science and industrial park, which speeded up MTK's commercialisation process and increased their marketing opportunities; 3) mainland China, especially in the Shenzhen area, had strong manufacturing capability. As more and more Taiwan companies entered the mainland China market, the close manufacturing network was formed and ready for MTK's enablement; 4) MTK, as the typical Taiwan company, was able to cut down the price hugely.

VCD chips had two groups: front end and back end. The front-end chips dealt with optical reading and control, and in 1996 the standard design included seven chips. The back end (MPEG) was the video decoder chip (search for the competitive strength 2007). MTK integrated these two groups of chips into two chips, and finally into a single chip by continuous technology improvement. Furthermore, MTK also increased the reading speed of the VCD to 24X, which was much faster than that of mainstream 4X or 8X. Finally, MTK overcame all the competitors and dominated this market.

At the same time, DVDs were also becoming popular. MTK brought the same strategy and methods into this market in 1999. One of the key principles held by MTK was that they preferred incremental innovation based on their existing technology capability, at the ratio of 3:7. The DVD technology only added a few technologies based on the VCD technology. For example, the video decoder was upgraded into MPEG2 from MPEG1. As the dominator of the VCD market, MTK successfully brought their advantages to the DVD market even though the market was not yet mature.

MTK's behaviour broke the rule of control by Western and Japanese companies. MTK made the industry more specialised and standardised by integrating the R&D part. This trend encouraged downstream partners, and stimulated their passion for innovation. MTK did much to contribute to the industry's production efficiency and effectiveness.

In summary, this project helped MTK to shape its typical strategy: first, within its limitations of basic research and development, MTK would penetrate a mature industry by providing an easy to adopt, low-priced Turnkey solution. Thus, this solution would attract many partners to work together. Second, it was key to find the right partners to use this Turnkey model. MTK was lucky as many of its partners in the Hsinchu science park already had business in mainland China, especially in southern China. So, MTK could enble the success of their partners' network as well as trigger a local manufacturing network. The local business ecosystem helped MTK scale up their shipment and win first place in the VCD and DVD market. This project highlights the importance of the business ecosystem around MTK. Third, MTK also quickly upgraded its product into relevant categories where it used the same strategy. For example, DVD was not a purely brand-new product but was an upgrade from VCD. MTK carried their VCD market strategy into the DVD market.

7.3   Project 3b: Mobile 2G

MTK became a famous star in 2008 for nurturing China's booming local mobile phone market. The Turnkey solution was already used by more than half the Chinese local mobile phone market. This kind of idea renewed the mature mobile industry into another low-cost mobile phone, which was a huge niche market. MTK also led an evolving view of their Turnkey solution. The mobile phone value chain contains the chip design, system design, manufacturing, industrial design and mechanical design, sales and after-sale activity. Top design companies like TI and Infineon focused on chip design. Top OEMs and Taiwan OEMs focused on system design, logistics and manufacturing. Mainland China OEMs focused on manufacturing. White brand OEMs only focused on sales and assembly. MTK started its mobile chip department in 2001 and developed their first chips in 2004. However, the top OEMs, even Taiwan OEMs, refused to adopt MTK's chip platform. At that time, Taiwan OEMs like BenQ adopted TI's solution which took as long as nine months to develop new systems based on TI's chips. In order to shorten the lead time, MTK tried to persuade local OEMs to adopt MTK's Turnkey solution. Finally, only one small OEM adopted MTK's solution because at that time MTK did not have a reputation for mobile chips.

Besides, at that time, the mobile phone market in mainland China was ramped up. MTK discovered that, in mainland China, there was no high design-capacity company but there were many manufacturing-related companies as well as the huge market. Learning from our previous success on VCD and DVD, they started to introduce the Turnkey solution to the manufacturing network in mainland China in order to enable their innovations.

7.3.1 Redesign the Turnkey solution

They redesigned the Turnkey model. MTK integrated the RF chip, baseband chip and power management chip into one chip as well as embedding an operating system and some basic application software. MTK was further deeply engaged in helping local partners to adopt MTK's solution. This Turnkey model left downstream design companies with very little system design. Meanwhile, industrial design, manufacturing and sales were the key competitive advantages of mainland China players.

7.3.2 MTK and its business ecosystem

MTK also realised the importance of partners' collaboration in its business ecosystem. They supported many IDHs (independent design house) which could deliver system design solutions based on MTK's Turnkey model. This system design allowed downstream players to easily develop new mobile phones. MTK also provided many training sessions on MTK's technical specification, which made more engineers familiar with MTK's solution.

MTK had its complementary supporting manufacturing network – Shanzhai network (Zhu & Shi 2010). Within the network, the key role was Shanzhai OEMs (white brand OEMs) which coordinated the whole manufacturing system's cooperation and achieved quick response to the market. The Shanzhai network included the IC design company, ID&MD (industrial design and mechanical design) company, Casing factory, distributors as well as EMS companies. First, the IDH got the IC solution from IC design companies like MTK, and then provided mobile system designs to OEMs. OEMs then brought the market information from franchised stores to IDHs and also coordinated with the IDH, ID&MD companies and Casing factories to finalise the mobile design. Then with help from distributors, OEMs got all the bills of material, and EMS would assemble the parts and test these mobiles. These mobiles would be delivered by different levels of franchises to customers. With the low-price Turnkey model and quick response manufacturing, all these manufacturing processes only took around 21 to 40 days from OEMs' perspective. The price usually varied from only 20 to 80 British pounds.

The Shanzhai mobile phones were divided into two categories: first, imitating the design and function of branded cell phones but with much lower price; second, making cell phones of more creative appearance and with unique functions. With the unique interface and unbelievably low price, Shanzhai mobiles were very popular. They could be upgraded with significant functions very quickly by upgrading the Turnkey solution.

7.3.3   Upgrading the Turnkey solution

MTK also continued to improve the performance of their Turnkey solution in order to cope with increasing requirements. Focusing on integration of the Turnkey model from the baseband chips of MT6205 to MT 6209, these chips enabled the China mobile manufacturing network to boom as the so-called 'Shanzhai network'. Shanzhai originally meant grass-roots power. The Shanzhai network is demonstrated as a cluster network formed by small private companies in the Shenzhen area.

The chips from MT6205 to MT6229 were all baseband chips which were the core data processing chips. MTK tried to integrate all functions into a baseband chip as Figure 7.4. They added MP3 and Wap, GPRS into MT6218, Camera and MP4 into MT6219, online camera to MT6226, TV

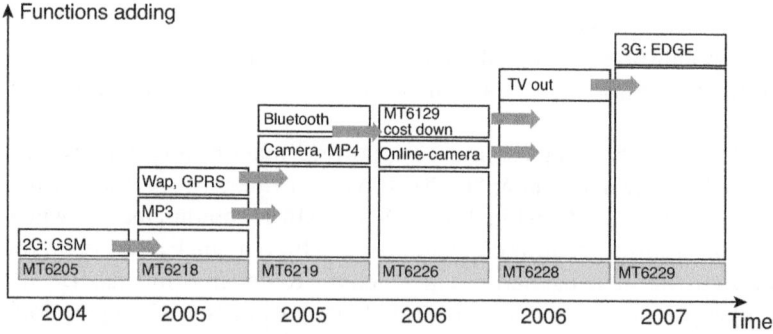

Figure 7.4   Turnkey solution development

function to MT6228 and 3G function to MT6229. MTK also designed an RF chip (for receiving and sending signals) and a power management chip. MTK combined these three chips (baseband, RF and power management chips) together with software to be the so-called Turnkey model.

In summary, this project further tested the MTK's previous strategies by adding more recognition of the operation of a business ecosystem. The business ecosystem partners helped MTK differentiate the final mobile phone design and finally helped hugely to scale up its production.

7.4   Project 3c: Smartphone

The mobile computing industry emerged with the Mobile 3G era. There are three standards in the current 3G market: TD-SCDMA from China, CDMA2000 from Qualcomm of USA and WCDMA from Europe.[3] In the China market, there are only three state-run mobile carriers licensing each of these three 3G standards. China Mobile has the license of TD-SCDMA; China Unicom licensed CDMA2000 from Qualcomm; China Telecom licensed WCDMA from its European side.[4]

In the current 3G mobile computing market, one of the typical products is the smartphone, an end-user device that requires functions varying from entertainment to computing. In this industry, people focus more on the customised service than on the simple phone service. For example, carriers act not only as the signal channel landlord as in the mobile 2G industry, but also as service providers in the smartphone market by providing many applications. In conclusion, the mobile computing industry focuses not only on device manufacturing but also on customer service. Faced by these changes from the mobile 2G application, MTK began to change from just a Turnkey solution provider into a more active role to co-evolve with partners. To summarise, there are three ways that MTK adapted:

First, in terms of hardware, MTK integrated different industry standards with its Turnkey model in order to cut down the cost and entry barriers. There were three main standards for 3G mobile (TD-SCDMA, WCDMA, CDMA2000). As MTK realised the emergence of 3G mobile, they began to deploy our positions in those three standards very early. For example they acquired ADI and got TD-SCDMA IP license – a Marketing Manager of MTK. Later, they cooperated closely with Qualcomm (the number one chip design company) in order to get a part license for WCDMA, most of the CDMA2000 license was owned by Qualcomm and it was too expensive to license from them. After that, MTK integrated each license and proposed a new Turnkey model specifically for the smartphone market.

Second, instead of just embedding an operating system as in the mobile 2G Turnkey model, MTK supported an open-source OS – Android – in the smartphone market. The Android community was a big group of leading firms including carriers and OEMs.[5] Partners developed an Android-based operating system and application software which was compatible with MTK's new solution. Specifically, MTK also co-evolved with many ISVs by embedding their software. For example, QQ, developed by Tecent, was the most popular online real-time communication tool. Before 2009, each Shanzhai mobile phone with QQ tools was designed by the IDH, not by MTK itself. MTK found that almost every IDH wanted to embed QQ tools into MTK's Turnkey mode, so in order to reduce lead time, MTK embedded QQ into its platform. Then, MTK launched a cooperation with Tecent in the hope that Tecent would develop QQ based on MTK's platform. At the same time, MTK also provided a software development kit for software partners. As MTK's solution was the most popular after 2007, Tecent was very happy to work with MTK. Also, MTK was continuously improving its hardware design: the MT6226 model could support online video, based on QQ tools. It was a double win for both MTK and Tecent. Besides this project, MTK also hoped to persuade Microsoft to offer a better price for embedding the Windows mobile operating system.

Third, in the 3G era, carriers would act as the platform for customer service and feedback. In order to expand its market share, MTK also tried its best to connect with carriers by persuading them to adopt its solution for 3G mobile phone development. MTK tried to persuade China Mobile (the number one carrier in China)[6] and then they both joined the group for a TD-SCDMA alliance in 2008.

In summary, MTK further enriched their three typical strategies: the Turnkey model; the business ecosystem strategy; and upgrading technology in relevant products, especially on the business ecosystem side: MTK kept improving technology from the mobile 2G and finally entered the mobile 3G market. As smartphones were much more complicated than the 2G mobile phone, MTK acted more actively in the smartphone market than in the mobile 2G industry. They had already realised the importance of co-evolution with partners in the mobile computing industry. In its business ecosystem, MTK not only cooperated with hardware and software partners, but also expanded their partners' networks by adding carriers, open-source operating system and industrial associations. MTK also dedicated themselves to software development and network resource coordination.

Table 7.1 MTK's business ecosystem development by phase

Projects	Business ecosystem evolution				
3a–VCD/ DVD	1. Single chip solution; cut down high technology barrier and low cost	1. Support various partners from Hsinchu park	1. Follow Taiwan OEM 2. Enable Shenzhen manufacturing network	1. Keep low cost and low entry barrier 2. Improve solution into 24X	1. Upgrading to the DVD chip to renew industry
3b– Mobile 2G	1. Introduce mobile phone solution 2. collaborate with local Taiwan OEM	1. Improved the Turnkey model 2. Enable Shanzhai manufacturing network	1. Improved single-chip solution 2. Selected partners regarding their performance (IDH)	1. Continue to improve the Turnkey model 2. Low-cost and luxury mobile phone existed in market	1. MTK introduced low cost of Smartphone solution 2. MTK succeeded 2G partner network
3c– Smartphone		1. Introduce smartphone succeeding from 2G firm network 2. Collaborated with hardware partners 3. Co-design solution with local company 4. Design support and training 5. Collaborate with carriers	1. Improved the Turnkey model 2. Selected partners regarding their performance (IC design, IDH)		
Phase status	1. New solution introduction 2. Initiate the new supply chain	1. Solution to attract more partners 2. Co-evolve with different partners	1. Collaborated with specific partners of Shanzhai network 2. Converged solution	1. Close relationship with partners 2. Solution continuous improvement	1. Introduce new idea 2. Succeeded previous network
Phases	Phase 1	Phase 2	Phase 3	Phase 4	Phase 5

7.5   MTK's evaluation and its business ecosystem development

MTK's three projects: the VCD/DVD, mobile 2G and smartphone also demonstrate the five phases of the BELC as shown in Table 7.1.

In the first phase, MTK introduced its Turnkey solutions (VCD, Mobile 2G), and was trying to persuade more partners to adopt its solutions. In the second phase, MTK joined an open-source community and co-evolved with local ISV (independent software vendors) by adopting local requirements. MTK also provided training sessions so as to create a community of MTK chips-based design groups.

In the third phase, MTK provided the improved Turnkey model with low-cost and low-entry barrier to trigger innovation in the Shanzhai manufacturing network. This strategy was very successful in the VCD/DVD and mobile phone project. With this success, in the fourth phase, MTK approached dominance of design and maintained industry structure as well. In the last phase, niche ideas were introduced to renew the industry. For example, the DVD solution aimed to renew the VCD industry.

Part III

Theory Construction of Business Ecosystems

By conducting cross-cases analysis, Part III finally delivers the theory construction of business ecosystem including the life cycle, constructive elements, configuration pattern and nurturing process of business ecosystems

Part III

8
The Business Ecosystem Life Cycle and Its Phase-Ending Status

Learning from the three cases, by combining the common features of each phase, this chapter is to develop the typical life cycle of a business ecosystem. Moore's life cycle was mainly developed from a stable PC industry, while our data was collected from the emerging mobile computing with very dynamic and uncertain nature, which will update Moore's business ecosystem life cycle.

8.1 Five phases, identification

By cross-case analysis and combining the findings from ARM, Intel and MTK, the five-phase life cycle was identified from cross-case analysis as shown in Table 8.1.

In the first phase, a new solution is proposed for the emerging market. Then the central firm leverages partners to initiate the supply chain for the new market. After they produce the new solution with a simple supply chain, the life cycle enters the second phase. Phase one can be called 'Emerging' based on its features.

In the second phase, solution diversity is highly encouraged to meet the uncertain market. The partner network is very flexible with high interoperability. So this phase can be called 'Diversifying'.

In the third phase, with the solution selection process, the market becomes specialised and converges on the solution. As a result, the partners' network becomes integrated and focused on those specialised markets. Phase three can be called 'Converging'.

The fourth phase may last for a long time as the dominant design is approached. The partners' network is stable and forms a close alliance for mass production of that dominant design. So this phase is called 'Consolidating' because of its 'stable' feature.

Table 8.1 Cross-cases analysis for phase identification

Projects	Business ecosystem evolution				
Learning from ARM table in Chapter 5	1. New solution for Emerging industry 2. Simple supply chain initiated	1. ARM IP made industry solution diversity 2. Industry highly collaboration and flexible connection	1. Market specialisation and convergence 2. Supply chain finalised	1. Solution continue improving 2. Supporting fixed relationship	1. Solution upgraded
Learning from Intel table in Chapter 6	1. Introduce new solution for emerging market 2. Simple supply chain	1. Partly open hardware 2. Software partner involvement around Atom-based project	1. Market specialisation and solution convergence 2. Partner network converged	1. Dominant design 2. Fixed partners 3. Stable partner network structure	1. New/niche market2. Project failure and sold out
Learning from MTK table in Chapter 7	1. New solution introduction 2. Initiate the new supply chain	1. Solution to attract more partners 2. Co-evolve with different partners	1. Collaborated with specific partners of Shanzhai network 2. Converged solution	1. Close relationship with partners 2. Solution continuous improvement	1. Introduce new idea 2. Succeeded previous network
Cross-cases analysis	1. New solution for emerging market 2. Simple supply chain	1. Solution diversity 2. Partners' network highly interaction	1. Market specialisation and solution convergence 2. Partner network converged	1. Dominant design 2. Stable partner network structure	1. New/ niche market 2. Partner network reorganisation or Industry decline
Phases	Phase 1: Emerging	Phase 2: Diversifying	Phase 3: Converging	Phase 4: Consolidating	Phase 5: Renewing

In the last phase, niche markets are emerging, and the partners' network starts to reorganise itself in order to renew the original market. The original market can be replaced by the emerging market or may experience recession for another long period. If the original market is replaced by the emerging market, firms will re-enter phase one to repeat the five phases. As a result, this phase can be called 'Renewing'.

In summary, the BELC has five sequential phases: Emerging, Diversifying, Converging, Consolidating and Renewing.

8.2 Phase-ending status

In each phase there are also the phase starting and ending statuses. In order to clarify the phase boundary, those statuses also need identifications (the ending status of one phase is also the starting status of the next phase). By using Moore's model (Moore 1996) and to Table 8.1, four levels of components will be selected in order to identify each phase status: products, core firms, partners' network and the rest of the business ecosystem in Figure 8.1.

In Status 0 (Pre-emerging), there are novel ideas proposed by core firms. However, at that moment, partners are separate and not organised.

In Status 1 (Post-emerging), the core firms coordinate a simple supply chain and finally initiate the products. The partners' network is formed as a single supply chain.

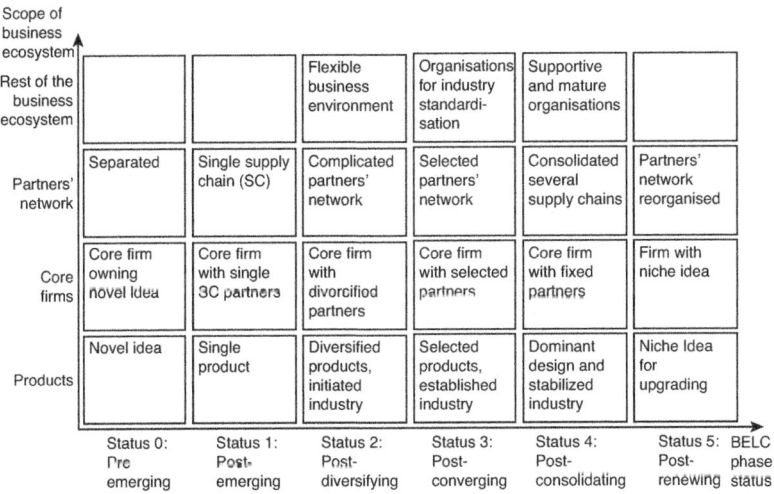

Scope of business ecosystem						
Rest of the business ecosystem			Flexible business environment	Organisations for industry standardisation	Supportive and mature organisations	
Partners' network	Separated	Single supply chain (SC)	Complicated partners' network	Selected partners' network	Consolidated several supply chains	Partners' network reorganised
Core firms	Core firm owning novel Idea	Core firm with single SC partners	Core firm with divorcified partners	Core firm with selected partners	Core firm with fixed partners	Firm with niche idea
Products	Novel idea	Single product	Diversified products, initiated industry	Selected products, established industry	Dominant design and stabilized industry	Niche Idea for upgrading
	Status 0: Pre emerging	Status 1: Post-emerging	Status 2: Post-diversifying	Status 3: Post-converging	Status 4: Post-consolidating	Status 5: Post-renewing	BELC phase status

Figure 8.1 Phase-ending status of the business ecosystem life cycle

In Status 2 (Post-diversifying), core firms and their diversified partners make the product solution very diversified to meet dynamic business requirements. After the phase of Diversifying, products have become diversified, and the partners' network has become flexible and complicated. Also non-direct business partners (the rest of the business ecosystem) have become flexible to support activities for enabling diversified solutions. As a result, a potential new industry has begun.

In Status 3 (Post-converging), some product solutions are selected to meet specialised market requirements. Many partners are also selected to form related supply chains. Other organisations also provide industry standardisation. As a result, the new industry is established.

In Status 4 (Post-consolidating), the dominant design has already been proposed in order to improve industry productivity. The partners' network is organised as a fixed supply chain. As a result, the industry is very stable. Also, the rest of the business ecosystem can be very supportive and mature for future industry development.

In Status 5 (Post-renewing), if new ideas successfully upgrade the existing industry, this status is exactly the same as Status 0. Another life cycle will then begin. If new ideas cannot upgrade the existing industry, this status will be the recession of the industry.

With learning from the three main case studies and their nine projects, the BELC is proposed as having five phases: Emerging, Diversifying, Converging, Consolidating and Renewing. Figure 8.2 presents the

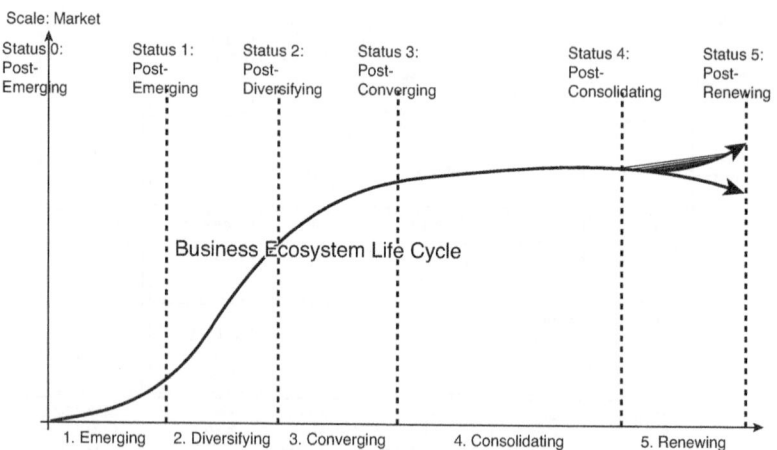

Figure 8.2   The business ecosystem life cycle and phase status

sequential phases, phase-ending status (connecting between phases) (horizontal view) and market scale (vertical view).

8.3   Discussion of life-cycle study

There are many life-cycle concepts such as the product life cycle (PLC), technology life cycle (TLC), firm life cycle, industry life cycle (ILC) and business ecosystem life cycle (BELC). Table 8.2 makes the comparison between those different life-cycle studies.

PLC based on the evolution of sales consists of four sequential stages: introduction, growth, maturity and decline with an S-curve shape (Levitt 1981; Polli & Cook 1969). PLC links with Roger's innovation diffusion theory, as product scale growth indicates the way innovation and technology are adopted. It demonstrates the key relationship between

Table 8.2   Comparison between different life-cycle studies

	Perspective	Stages and content	Main purpose and their relationship
Product life cycle	Product/ application	Introduction, growth, maturity and decline	To understand product from a dynamic perspective to forecast market sales
Technology life cycle	Technology	Technology development, technology application, application launch, application growth, technology maturity and degraded technology	Understand technology penetration and time to technology transfer internationally
Firm growth life cycle	Firm growth	Many types of stages: fitting structure, system pattern and management practices	Understand firm growth and its relevant strategy
Industry life cycle	Industry	An early exploratory stage, an intermediate development stage and a mature stage	Under relationship between ILC and TLC, PLC and firm growth
BELC	Business ecosystem	Emerging, Diversifying, Converging, Consolidating and Renewing	Cross-industries: to cultivate business environment including partners and ideas for future product development

innovation and industry maturity (Rogers 1962). TLC reflects the PLC with six similar stages: technology development, technology application, application launch, application growth, technology maturity and degraded technology (Harvey 1984). Furthermore, scholars also analyse the extension of TLC through technology transfer from developed countries to less developed countries. Phase five of technology maturity is the right point at which to transfer the mature technology to less developed countries, and to get mutual benefit (Vernon 1966; Harvey 1984).

Firm life cycle outlines the key activities for different stages in a firm's growth, specifically on its fitting structure, system pattern and management practices (Garnsey 1998; Greiner 1997; Churchill & Lewis 1983; Rutherford et al. 2003). In terms of ILC, Porter (1980) thought that even product sales had a big impact on industry growth, and ILC does not always go through the S-shaped pattern of PLC. Sometimes ILC would skip some phases of PLC (Porter 1980). ILC could be divided into three stages: the early exploratory stage, intermediate development stage and mature stage (Williamson 1975). More specifically, ILC studies focused on the number of firms and the nature of innovation activities (Audretsch & Feldman 1996). Peltoniemi also argued that ILC studies do not address the interconnectedness of industries and how these would affect ILCs. He argues that the inter-industry experience would generate a mature industry which would lead to the birth of another new industry (Peltoniemi 2011).

In summary, the TLC, the PLC and the firm growth life cycle all reflect the ILC, where technologies are adopted; product and process innovations are implemented; firms enter or exit the current industry. As a result, they are naturally connected.

The BELC was first proposed by Moore in 1993 as birth, expansion, authorities and renewal by highlighting the interaction among firms, industries and their business environment (Moore 1993, 1996). Moore expanded the previous life-cycle scope of firms, product sales and innovation activities to cross industry boundaries and to include different levels of organisations such as government agencies, industry associations, competitors and business opportunities. This study enriches the BELC phases: Emerging, Diversifying, Converging, Consolidating and Renewing in order to cope with an emerging dynamic business environment. This is because Moore's work focused on the competitive advantage development by the activities in the expansion and authorities phases which ignored dealing with the dynamic features and uncertain future of the market. This chapter argues that diversification strategy should highlight coping with the future dynamic market. Diversified products, rather than the dominant design, should be encouraged to meet various market requirements. The detail of the nurturing process from the firm perspective of each phase will be explained in the next chapter of the business ecosystem nurturing process.

8.4   Conclusion

This chapter integrated those nine different projects as case studies of three companies. ARM provides its development in the projects of mobile phone business, leader partner strategy and IP categorisation; Intel provides its development in the projects of PC industry business, Xsacle and Atom; MTK provides its development in the projects of VCD/DVD, mobile 2G and smartphone.

By cross-case analysis of these projects, this chapter proposed that the BELC has five phases of Emerging, Diversifying, Converging, Consolidating and Renewing, each phase with its own ending status. Different phases demonstrate different business contexts which can have impacts on firms' strategy and behaviour.

Then a discussion was conducted over the various life-cycle studies such as the PLC, TLC, firm growth, ILC and BELC in order to highlight the contributions by the BELC.

Learning from the main case studies, data from these three main cases was presented by the BELC phases in the Appendix Figure 1 (ARM), Appendix Figure 2 (Intel), Appendix Figure 3 (MTK). Following the Appendix figures, the next three chapters will study the business ecosystem constructs, configuration patterns and nurturing process in the context of the five phases of the business ecosystem life cycle.

9
Business Ecosystem Constructive Elements

9.1 Introduction

As learned from the review of existing literature, the constructs of a business ecosystem should be addressed further as the system could be understood through exploring its constructive elements (Von Bertalanffy 1969). Accordingly, many scholars applied this constructs study to different levels of manufacturing systems: in 1984, Hayes and Wheelwright highlighted constructs study by the framework of 'structure–infrastructure' model since constructive elements had a big impact on system manufacturing strategy (Hayes & Wheelwright 1984). In 1998, Shi and Gregory applied this framework at the international manufacturing system level as intra-firm level (Shi & Gregory 1998) and Harland addressed the inter-firm supply-chain level (Harland 1996) as well. Other scholars also paid attention to global engineering networks (Zhang et al. 2007) and global supply network level (Srai & Gregory 2008). As a result, this chapter has adopted the framework of 'structure–infrastructure' to deconstruct the business ecosystem in order to understand the business ecosystem itself.

This chapter aims to identify the constructive elements through learning about the life-cycle phase statuses from status 0 to status 5. The constructive elements of a business ecosystem would have a different impact on each phase and then generate the ending status of each phase. The general outcomes of this chapter are as follows:

- To identify the key constructive elements relevant to each phase status of the business ecosystem life cycle (BELC);
- To develop the generalised constructive elements of a business ecosystem.

9.2   Data analysis

9.2.1   Background and preliminary findings

General system theory (input-system-output) (Von Bertalanffy 1969) and 'structure–infrastructure' framework (Hayes & Wheelwright 1984) will be used as guidelines to deliver the constructive elements of a business ecosystem.

The 'structure–infrastructure' framework (Hayes & Wheelwright 1984) was proposed in order to understand a system more comprehensively. On the structural side, the elements are used to indicate the stabilised features of a system and the system activities that have a long-term orientation, while on the infrastructural side the elements are used to demonstrate system changes and the system activities with a short-term orientation.

As a business ecosystem has a life cycle, this chapter also addresses the difference among the five sequential phases that the constructive elements perform in: Emerging, Diversifying, Converging, Consolidating and Renewing. As the constructive elements are static, they can be learned from the ending status of the different phases which directly reflect different performances of the constructive elements.

As a result, two steps are followed in order to identify the constructive elements: first, the specific constructive elements of each phase status will be identified; then, by comparing those phases, the generalised constructive elements of a business ecosystem will be identified.

9.2.2   Constructive elements findings from each main case

This section aims to perform a cross-case analysis in order to enrich the findings from the literature. Taken from Appendix Figures 1, 2, 3, three mains cases (including nine sub-cases) are placed in Tables 9.1–9.4. In each case study, all the activities are divided into structure orientation and infrastructure orientation, which could be further mapped as structure and infrastructure elements. Specifically, the nature of structural elements and activities is that they are of long-term impact and difficult to reverse, while the nature of infrastructural elements and activities is that they are of short-term impact, usually specific operating aspects of business (Hayes & Wheelwright 1984; Harland 1996; Shi & Gregory 1998).

In each table the horizontal view indicates six sequential phase statuses along BELC while the vertical view demonstrates the three sub-cases and their activities with structure and infrastructure orientation.

The phase statuses and how they relate to different projects were developed from Appendix Figures 1, 2, 3. Some special instances should be highlighted: 1) the status 0 of Pre-emerging is identified from the project background; 2) in ARM projects, ARM7 (1a) was recycled into an embedded market, so the renewing phase of mobile/ARM7 project was just the emerging phase of the embedded market (1b-leader partners strategy); 3) in MTK projects, the mobile 2G (3b) project successfully renewed itself into the smartphone project. As a result, the renewing phase of mobile 2G project was also the emerging phase of the smartphone project (3c).

In summary, the whole table will describe the three project activities of each firm by the phase status of BELC. By using cross-case analysis in each phase, at the bottom of each table each case will deliver structural and infrastructural elements of business ecosystem by phase.

9.2.3   Cross-case analysis process

The conclusions from Tables 9.1–9.4 are placed into Table 9.6 in order to synthesise three case findings and deliver constructive elements by phase. Table 9.6's horizontal dimension is six-phase status while the vertical dimension presents three case findings with structure and infrastructure orientation. This table makes cross-case analysis and then finalises key influential constructive elements with the following three steps:

In the first step, the conclusion parts of each of the three cases are divided into either structural or infrastructural.

In the second step, cross-case comparisons of the phase status elements are conducted in order to draw conclusions about the elements.

In the third step, the second step is repeated regarding the six phase statuses, as constructive elements have played different roles in each phase status.

Finally, the details of the constructive elements of each phase status are presented in the following section.

9.3   Constructive elements by phase

Table 9.5 generated the constructive elements of a business ecosystem by phase, through synthesising single case analysis (Tables 9.1–9.4). The structure and infrastructure orientation activities are distinguished and the constructive elements reflected from the activities are also identified. These constructive elements also reflect each phase-ending status. In each phase, different constructive elements would have strong or

Table 9.1 Constructive elements from the analysis of ARM case

Projects	Orientation	Status 0: Pre-emerging	Status 1: Post-emerging	Status 2: Post-diversifying	Status 3: Post-converging	Status 4: Post-consolidating	Status 5: Post-renewing
1a-Mobile	Structure orientation	1. Separated partners' network	1. New IP solution 2. OEM pull: help from Nokia	1. Leader partners as TI 2. Interaction around ARM IP platform 3. IP evolution with leader partners (TI) on performance, capability	1. ARM based chip winning market advantage 2. TI's snapdragon platform winning the competitive advantage	1. ARM IP continue improving.	1. ARM7 redesigned for embedded market
	Infrastructure orientation	1. Idea for low power IP	1. Search future market 2. Persuading TI with help; 3. Expansion relationship; 4. ARM9 continually developed by TI	1. 3rd software partners supported by ARM 2. Tool development based on new OS for application developer	1. Strong alliance with NOKIA and TI	1. ARM support different IC platform with more activities	1. New partners found for the IP 2. Niche market emerging
1c-Leader Partner Strategy	Structure orientation			1. Leader partners identified; 2. ARM platform 3. Solution diversity based on ARM's platform	1. More partner become leader partner strategy (top players and potential players) 2. Tool support	1. More than 1000 customers chosen STM32 (Cortex-based) 2. Luminary micro chips win market 3. Consolidated design with LP	
	Infrastructure orientation			1. Huge embedded market and IP architecture diversity 2. Leader partners co-evolved 3. Industry vision co-designed	1. Huge embedded market and IP architecture converged 2. Small alliance formed by ARM and leader partners 3. Strong software partners	1. Strong software ecosystem 2. Market finalised	

Continued

Table 9.1 Continued

Projects	Orientation	Status 0: Pre-emerging	Status 1: Post-emerging	Status 2: Post-diversifying	Status 3: Post-converging	Status 4: Post-consolidating	Status 5: Post-renewing
1e-IP category:	Structure orientation				1. Leader partners strategy widely used 2. Feedback from partners 3. Succeeding from ARM7 ecosystem 4. Three streams of IP solution	1. Three streams to market segment 2. Dominant design approached	
	Infrastructure orientation				1. Market specialization 2. Encourage partners contribute to ARM platform	1. Partners reorganised into three business ecosystems	
Preliminary findings on constructive elements from ARM	**Structure**	1. Separated partners	1. Simple products 2. Existing resource like Nokia's network 3. Role: key OEM; key buyer, central firm 4. Platform based partner network	1. Products co-designed 2. Solution platform interacted 3. Role: leader partner, OSV, ISV and other supporting role	1. Integrated Solution platform 2. Selected product 3. Role: specific partners	1. Solution finalised 2. Platform stabilised	1. New niche IPs 2. Partners reorganised
	Infra-structure	1. Idea for future	1. Relevant market 2. Simple supply chain for new product 3. Partner leveraged by ARM	1. Flexible networking 2. Encouragement of participation 3. Co-created vision 4. Market uncertainty	1. Selected relationship 2. Market specialised 3. Close alliance	1. Stabilised network partners 2. Market stabilised	1. Niche market

Table 9.2 Constructive elements from the analysis of Intel case

Projects	Orientation	Status 0: Pre-emerging	Status 1: Post-emerging	Status 2: Post-diversifying	Status 3: Post-converging	Status 4: Post-consolidating	Status 5: Post-renewing
2a-PC	Structure orientation		1. Intel with its X86 processor 2. Industry has its vertical integration model	1. Intel processor support several OEMs platform;	1. PCI standard introduced by Intel	1. More standards introduced to maintain advantage	1. Strong ARM chips used to enter mobile
	Infrastructure orientation	1. Idea of focusing on processor area	1. Semiconductor industry specialised	1. Horizontal model enable various partners interaction	1. Attract partners around PCI platform	1. Close alliance around Intel and Microsoft	1. Partners persuaded to enter mobile market
2b-Xscale	Structure orientation	1. ARM license got from a lawsuit case	1. Penetrate into mobile by using ARM IP		1. Collaboration with Motorola-Moto A1200 handset	1. Manufacturing part outsourced to TSMC	1. Xscale chip abandoned
	Infrastructure orientation	1. The idea to enter mobile phone market	1. Market identification 2. Low ecosystem support		1. High manufacturing and maintenance cost	1. Competitors with strong partners support 2. Market dominated by ARM IP	1. Challenges from AMD to Intel PC processor
2b-Atom	Structure orientation	1. EeePC – a good-enough solution of notebook 2. Partners from PC industry	1. Learning from EeePC idea 2. Atom processor 3. New concept of Netbook and MID 4. Moblin ecosystem	1. Atom co-evolving with partners 2. Previous partners involvement and enable the diversity 3. Open project of operating system	1. Netbook standards with partners finalised		

Continued

Table 9.2 Continued

Projects	Orientation	Status 0: Pre-emerging	Status 1: Post-emerging	Status 2: Post-diversifying	Status 3: Post-converging	Status 4: Post-consolidating	Status 5: Post-renewing
Preliminary findings on Constructive Elements from Intel	Infrastructure orientation	1. Idea for low power chips 2. Market emerging	1. Previous partners succeeded 2. Application partners supported around Atom processor 3. Concept of Netbook	1. Partners' involvement 2. Moblin ecosystem support 3. Existing partners collaborated for MID product: Tecent, MSN	1. Partners selected to stabilise Netbook production 2. Low Market identification		
	Structure	1. New idea 2. Previous partners	1. Products and new solution 2. Roles: core firm, former PC partner 3. Previous resource in PC industry	1. Open Solution platform: Moblin 2. Various product: OS, application software 3. Role: key OEM, Moblin partners OSV, ISV, previous PC partners 4. Resource searching for expanding market	1. Integrated Solution platform (PCI, StrongARM, Netbook platform) 2. Key OEM, central firm	1. Product with more standard and improvement	1. Niche product 2. Role: key OEM, central firm, previous PC partners
	Infrastructure	1. Idea development for new market	1. Existing partners coordinated 2. New business direction development 3. Value chain creation 4. Market emerging	1. Supporting relationship 2. Future direction co-designed 3. Diversified market requirement	1. Market converged 2. Close networking	1. Close alliance	1. Niche and relevant market

Table 9.3 Constructive elements from the analysis of MTK case

Projects	Orientation	Status 0: Pre-emerging	Status 1: Post-emerging	Status 2: Post-diversifying	Status 3: Post-converging	Status 4: Post-consolidating	Status 5: Post-renewing
3a-VCD /DVD	Structure orientation	1. Future product as VCD 2. Previous technology background	1. Single chip solution 2. Partners in Hshinchu park	1. Collaborating with OEMs within the park 2. Software development kit provided	1. Single chip solution 2. Work with VCD player and PC players	1. Improved Turnkey solution 2. VCD of China manufacturing network	1. The VCD industry renewed by DVD
	Infrastructure orientation	1. Market dominated by top players	1. Supply chain built up in Hschinchu park	1. To mainland China by following OEM	1. Innovation enabled in China 2. Market maturity: analyse the market maturity S-curve	1. High technology barrier 2. High technology barrier cut down and low cost 3. Local China manufacturing network	1. Relevant new industry identified 2. Market at early stage but huge amount
3b-Mobile 2G	Structure orientation	1. 70% original technology 30% innovation 2. Low cost mobile phone 3. Hshinchu park	1. Turnkey solution for mobile phone 2. Taiwan OEM cooperation 3. Hshinchu park	1. Following VCD and DVD strategy, finally enabling Shanzhai network in Shenzhen 2. Shenzhen manufacturing network 3. Turnkey model cutting entry barrier	1. Turnkey model cutting down entry barrier 2. Shenzhen local strong manufacturing capability 3. Supporting local design house	1. Turnkey model continually improved 2. Dependence on Shanzhai manufacturing network	1. New idea: more function required 2. Succeeding from 2G mobile phone network
	Infrastructure orientation	1. Partners in other industries	1. Collaboration with small OEM, not BenQ 2. Market dominated by top players, very mature	1. Flexible network coordinated by Shanzhai OEM 2. Fast innovation and short lead time 3. Design support 4. Very flexible social system and policy regulation	1. Down-stream innovation fast enabled in Shanzhai network of Shenzhen area 2. Market Agency governance 3. Frugal mobile phone delivered] 4. Flexible policy and social system	1. Close and flexible network with quick response and highly innovating 2. Flexible societal system 3. Policy became strict	1. Vision for next generation of mobile phone 2. Market emerging

Continued

Table 9.3 Continued

Projects	Orientation	Status 0: Pre-emerging	Status 1: Post-emerging	Status 2: Post-diversifying	Status 3: Post-converging	Status 4: Post-consolidating	Status 5: Post-renewing
3c-Smartphone	Structure orientation	1. New idea: more function required 2. Succeeding from 2G mobile phone network		1. Succeeding from 2G firm network 2. Encouragement with Smartphone produce	1. Market focused 2. Hardware partners selected for 3G standards 3. IPs continually improved		
	Infrastructure orientation	1. Vision for next generation of mobile phone 2. Market emerging		1. Design support and training session 2. Chip development with Android groups 3. Connection with chip designer and carrier 4. Market diversity	1. 3G license adopted 2. Partners (IDH, IC design) selected to form supply chain		
Preliminary findings on Constructive Elements from MTK	**Structure**	1. Idea for emerging market 2. Previous partners	1. Central roles to initiate product with local OEM 2. Emerging product 3. Using Shanzhai network	1. Product co-designed 2. Central role interacting with local partners, ISV, OSV3 local manufacturing network	1. Central role to trigger Shanzhai network 2. Integrated Solution platform 3. Previous partners	1. Finalised Solution platform 2. Integrated resources	1. Niche product to renew industry 2. Reorganised relationship
	Infra-structure	1. New development vision 2. Emerging market	1. New concept for product 2. New industry divided by standards 3. Market feedback	1. Supporting local partners 2. Flexible networking 3. Low cost but functional product 4. Flexible societal and policy system	1. Selected Shanzhai network 2. Frugal mobile accepted 3. Market maturity 4. Flexible societal system	1. Close alliance with quick response 2. Policy became strict	1. New product identification 2. Market transition

Table 9.4 Cross cases analysis

Constructs	Main cases	Status 0: Pre-emerging	Status 1: Post-emerging	Status 2: Post-diversifying	Status 3: Post-converging	Status 4: Post-consolidating	Status 5: Post-renewing
Structure	ARM	1. Separated partners	1. Simple products 2. Existing resource like Nokia's network 3. Role: key OEM; key buyer, central firm 4. Platform based partner network	1. Products co-designed 2. Solution platform interacted 3. Role: leader partner, OSV, ISV and other supporting role	1. Integrated Solution platform 2. Selected product 3. Role: specific partners	1. Solution finalised 2. Platform stabilised	1. New niche IPs 2. Partners reorganised
	Intel	1. New idea 2. Previous partners	1. Products and new solution 2. Roles: core firm, former PC partner 3. Previous resource in PC industry	1. Open Solution platform: Moblin 2. Various product: OS, application software 3. Role: key OEM, Moblin partners OSV, ISV, previous PC partners 4. Resource searching for expanding market	1. Integrated Solution platform (PCI, StrongARM, Netbook platform) 2. Key OEM, central firm	1. Product with more standard and improvement	1. Niche product 2. Role: key OEM, central firm, previous PC partners
	MTK	1. Idea for emerging market 2. Previous partners	1. Central roles to initiate product with local OEM 2. Emerging product 3. Using Shanzhai network	1. Product co-designed 2. Central role interacting with local partners, ISV, OSV3, local manufacturing network	1. Central role to trigger Shanzhai network 2. Integrated Solution platform. 3. Previous partners	1. Finalised Solution platform 2. Integrated resources	1. Niche product to renew industry 2. Reorganised relationship
	Conclusion	1. Separated roles 2. Niche idea	1. Role (central firms with value chain initiated) 2. Emerging solution 3. Extended existing resource	1. Roles (central roles with crossed value chain) 2. Diversified solution 3. Platform with easy access 4. Various resource combination	1. Roles (central firm with selected value chain) 2. Selected solution 3. Integrated Solution platform 4. Selected resources	1. Roles (central firm with integrated partners) 2. Dominant design 3. Solution platform continue improving 4. Integrated resource	1. Roles (central role to reorganise partners) 2. Niche idea

Continued

Table 9.4 Continued

Constructs	Main cases	Status 0: Pre-emerging	Status 1: Post-emerging	Status 2: Post-diversifying	Status 3: Post-converging	Status 4: Post-consolidating	Status 5: Post-renewing
Infrastructure	ARM	1. Idea for future	1. Relevant market 2. Simple supply chain for new product 3. Partner leveraged by ARM	1. Flexible networking 2. Encouragement of participation 3. Co-created vision 4. Market uncertainty	1. Selected relationship 2. Market specialised 3. Close alliance	1. Stabilised network partners 2. Market stabilised	1. Niche market
	Intel	1. Idea development for new market	1. Existing partners coordinated 2. New business direction development 3. Value chain creation 4. Market emerging	1. Supporting relationship 2. Future direction co-designed 3. Diversified market requirement	1. Market converged 2. Close networking	1. Close alliance	1. Niche and relevant market
	MTK	1. New development vision 2. Emerging market	1. New concept for product 2. New industry divided by standards 3. Market feedback	1. Supporting local partners 2. Flexible networking 3. Low cost but functional product 4. Flexible societal and policy system	1. Selected Shanzhai network 2. Frugal mobile product accepted 3. Market maturity 4. Flexible societal system	1. Close alliance with quick response 2. Policy became strict	1. New product identification 2. Market transition
	Conclusion	1. New vision initiated	1. Leveraged relationship 2. Common vision initiated 3. Platform-based simple value chain 4. Emerging market	1. Diversified relationship 2. Vision co-designed 3. Platform-based flexible network 4. Market diversity and flexible societal and policy system	1. Selected relationship 2. Finalised vision 3. Platform-based converging network 4. Market specialised and adaptable societal and policy system	1. Close relationship 2. Close network around platform 3. Supportive societal and mature policy system	1. New vision introduced 2. Market transition

weak impacts, so different elements can be highlighted in each phase. Table 9.5 is developed from Table 9.6. It is used to place some relevant elements on the same levels along six phase statuses in order to highlight the evolution of each element. This table also delivers the concept of each element along the life-cycle phases statuses.

9.3.1 Constructive elements of Status 0: pre-emerging

The Pre-emerging status demonstrates a firm starting to initiate new ideas for the emerging market requirement. There are many previous partners in the original industrial sectors around the key firm; however, they are not well organised (*Separated roles*) for the emerging industry. The firm at this moment is willing to bring a niche idea (*Niche idea*) to trigger the emerging industry development as well as to propose the future image (*New vision*) to encourage its previous partners and new partners to work together.

As a result, in Status 0, the key constructive elements are separated role and niche idea on the structural side, and new vision on the infrastructural side.

9.3.2 Constructive elements of Status 1: post-emerging

In the phase of Emerging, firms aim to commercialise the new solution for the emerging market by leveraging different partners. As a result, at the point of Post-emerging, first, the central firm has already provided the key emerging solution (*Emerging solution*) which would meet the future requirements of the emerging market (*Emerging market*). Second, in order to promote its solution, the central firm has already encouraged its partners (*Roles: central roles with supply chain initiated*) with the future development direction of the industry (*Common vision*) for this solution as well as approaching the existing resources (*Existing resources like partners, technology*). Thirdly, the central firm has also built up a supply chain around its solution (*Platform-based simple supply chain*) by leveraging its partners (*Leveraged relationship*), not only direct customers or suppliers but also all the players along the value chain.

As a result, in status 1, structural elements are: roles (central firm with value chain initiated), emerging solutions and extended existing resources; the infrastructural elements are: leveraged relationship, common vision, platform-based simple supply chain and emerging market.

9.3.3 Constructive elements of Status 2: post-diversifying

The phase of Diversifying enables market diversity along with diversified solutions. At the point of Post-diversifying, the image of the future industry is co-designed by partners (*Co-designed vision*). Normally, the solution platform from the central firm is easily accessed (*Solution platform with easy access*) by partners in order to enable them to contribute their part in order to make the product solutions diversified (*Diversified solution*). Different roles (*Role: central firm with crossed value chain*) are well organised to place themselves in different positions and complementary to one another (*Various resources combination*). This kind of network is very flexible and can re-organise partners (*Diversified relationship*) to contribute an increasing number of diversified ideas and also to make the nurturing process sustainable around the central firm's Solution platform (*Platform-based flexible network*). Also, the societal and policy systems are also very flexible to allow the market to be diversified (*Market diversity and flexible societal system and policy system*).

As a result, the constructive elements of Status 2 are structural: Roles (central firms with crossed value chain), Diversified solutions, Solution platform with each access, Various resources combination; and infrastructural: Diversified relationships, Vision Co-designed, Platform-based flexible network and Market diversity and flexible societal and policy system.

9.3.4   Constructive elements of Status 3: post-converging

The key idea found in the Converging Phase is used to build up an integrated Solution platform for several typical end-user applications. At the point of Post-converging, partners make common agreement on the future direction of the industry by finalising the future industry vision (*Finalised Vision*). The central firm with their Solution platforms could be selected (*Roles: central firm with selected value chain*) in order to meet *specialising* market requirements. More and more companies would integrate these platforms (*Integrated Solution platform*) with their own competences to deliver end-use products (*Selected solution*) which finally formulates the partner network converging around the central firm Solution platform (*Platform-based converging network*). Moreover, central firms and their selected partners (*Selected relationship*) are also searching for the right resources to support the expansion of their product volume (*Selected resources*) which would finally improve the industry efficiency and effectiveness. Furthermore, firms also use mature policy to drive the big development of a specific industry (*Market specialised and adaptable societal and policy system*).

The constructive elements at Status 3 are: structural: Roles (central firm with selected value chain), Selected solutions, integrated Solution platform and Selected resources; and infrastructural: Selected relationships,

Finalised vision, Platform-based converging network, Market specialised and adaptable societal system and policy.

9.3.5 Constructive elements of Status 4: post-consolidating

In the Consolidating Phase, central firms continue to introduce Solution platforms in order to maintain competitive advantage. As a result, at the point of Status 4, companies around central firms are more integrated in order to scale up the products volume (*Roles: central firm with integrated partners*). The dominant design (*Product as dominant design*) allows more partners to work together; as a result, the relationship is very stable among partners (*Close relationship*). The solution platform is continually improved by the central firm in order to maintain competitive advantage (*Continuing improvement of Solution platform*). The supporting resources are integrated by the central firm and its partners in order to improve industry efficiency (*Integrated resource*). After that, a very close partners' network is formed around the central firm's Solution platform (*Close network around platform*). Societal system and policy is also very adaptable to this partners' network (*Supportive societal and mature policy system*).

The constructive elements are: structural: Roles (central firms with integrated partners), Solution platform continue improvements and Integrated resources; and infrastructural: close relationships, finalised vision and supportive societal system and mature policy.

9.3.6 Constructive elements of Status 5: post-renewing

In the Renewing phase, the current market is saturated. Central firms introduce the vision (*New vision introduced*), which achieves the market transition from the original market (*Market transition*). This kind of idea (*Niche idea*) not only brings the new products but also brings the idea to change the partners' network (*Roles: central role to reorganise partners*), and the way they interact with one another.

The constructive elements in this phase are: structural: Roles (central firms to reorganise partners' network), Niche ideas; and infrastructural: new vision introduction and market transition.

9.4 General constructive elements of a business ecosystem

Table 9.5 is developed from Table 9.6 across the phases by placing the same type of constructive elements together. Table 9.5 presents those constructive elements that evolved along the six phase status. As a

Table 9.5 Concept of each constructive element along phase status

Constructs	Status 0: Pre-emerging	Status 1: Post-emerging	Status 2: Post-diversifying	Status 3: Post-converging	Status 4: Post-consolidating	Status 5: Post-renewing
Structure	1. **Separated roles**: the previous partners are separated without new targets 2. **Niche idea**: niche idea for the emerging market	1. **Role (central firms with value chain initiated)**: central firm formulates a supply chain for emerging solution 2. **Emerging solution**: simple product is proposed for the emerging industry 3. **Extended existing resource**: existing partners, technology	1. **Roles (central roles with crossed value chain)**: central firm works with different partners in the value network 2. **Diversified solution**: diversified idea and application for the growing market 3. **Platform with easy access**: central firm's Solution platform with easy access by partners 4. **Various resource combination**: various resource supporting central firm's Solution platform	1. **Roles(central firm with selected value chain)**: central firm works with selected partners 2. **Selected solution**: diversified products are selected to meet the specialised market 3. **Integrated solution platform**: alliance around central firm's Solution platform 4. **Selected resources**: resources match the requirement from central firm and its partners	1. **Roles (central firm with integrated partners)** : central firm and its fixed partner network 2. **Product as dominant design**: product with well-accepted market standard 3. **Continuing improvement of solution platform**: Solution platform continue improving to maintain its advantage 4. **Integrated resource**: resource is integrated by central firm and partners in order to improve industry efficiency	1. **Roles (central role to reorganise partners)**: partners without new target 2. **Niche idea**: niche idea is for emerging market

Infrastructure

1. **New vision initiated:** the vision is initiated for the emerging industry

Stage			
1. **Leveraged relationship:** central firm is to leverage partners besides direct suppliers and customers in order to build up supply chain	1. **Diversified relationship:** the relationship between firms is very flexible and of various combinations	1. **Selected relationship:** partners are selected by the central firms	1. **Close relationship:** stabilised relationship between central firms and partners
2. **Common vision initiated:** an anticipated industry future shared by partners	2. **Vision co-designed:** partners co-design the image of industry future	2. **Finalised vision:** partners finalised the image of industry	
3. **Platform-based simple value chain:** Core business process is implemented around central firm's platform	3. **Platform-based flexible network:** Core business process is flexible around central firm's platform	3. **Platform-based converging network:** Core business process is converged around central firm's platform	2. **Close network around platform:** Core business process is finalised around central firm's platform
4. **Emerging market:** the market emerged from relevant and existing markets	4. **Market diversity and flexible societal and policy system:** market is uncertain with various products	4. **Market specialised and adaptable societal and policy system:** market is specialised and policy has become mature	3. **Supportive societal and mature policy system:** the social system is very supportive to business activities

1. **New vision introduced:** vision is initiated for the emerging industry

2. **Market transition:** new market emerges from relevant and existing market

result, the generic constructive elements can be generated in Table 9.7. The definitions of each element are also presented in Table 9.8.

9.4.1　Structure perspective

Functional roles

In Status 0, Pre-emerging, different roles are separated as the central firm does not have tangible product and it is not ready to coordinate those roles. In Status 1, Post-emerging, the supply chain is set up by the central firm in order to commercialise its solution. In Status 2, Post-diversifying, different roles are formed in various value chains in order to commercialise the diversified ideas. In Status 3, Post-converging, some selected solutions are accepted by most of the partners, the position of the companies/roles will be finalised and different value chains will be formed which are crossed, complicated but well organised. Then, in Status 4, Post-consolidating, initiator, adopter and specialist are integrated to form a close alliance in order to maintain the competitive advantages for a long period. In Status 5, Post-renewing, the central firm brings niche ideas to penetrate into other relevant markets as the original market is saturated. At that time, the roles are separated.

Besides the changing roles along the life cycle, different roles have different operational strategies. For example, Intel is the dominator in the PC industry. However, due to the complicated context of the mobile computing industry, the dominator has begun to open its door to partners, which means the dominating behaviour is diminished in this emerging industry with a high requirement for interoperability. In this way the dominator's behaviour has changed to that of using their platform to stimulate partners' contribution. The role with these behaviours is like that of an initiator building up a new ecosystem. The other two main cases of ARM and MTK are also initiators as they initiate the business ecosystem around their Solution platforms. Some partners owning specialised capabilities contribute to this Solution platform; for example, many software partners provide operating systems and application software to enhance ARM IP's performance. Besides these partners, others adopted the platform to develop products; for example, Asus adopted Intel's chip to develop EeePC. As a result, roles can be specialised into three main types: initiator, specialist and adopter. Initiators are willing to build a business ecosystem with their platform and products; specialist firms are adding value to the initiator's platform; adopters are building up final products based on the initiator's and specialist's co-designed platform.

Furthermore, despite the strict distinction between roles' type, roles are more likely to adapt to match the purpose of each nurturing process. For example, ARM is regarded as the initiator with its IP platform, and Nokia acted as the adopter to embed ARM's IP into mobile phones; other software providers will be the specialists. In contrast, if Nokia is regarded as the initiator for the mobile phone platform, then the carrier can be seen as the adopter and ARM as a specialist. In this way, the three key roles are recognisable from the firm's perspective and a firm may act different roles in different firms' business ecosystems. As a result, firms keep changing their roles in order to adapt to the dynamic business environment. Thus, different roles will help a business ecosystem to become more professional and flexible.

Adaptive solution

In Moore's framework, the offer in business ecosystem is to deliver a package of rich and full core value (Moore 1996). In the case studies, some different points have been highlighted in Table 9.5. At Status 0, the central firm has already proposed the niche idea for the emerging market. Then, in Status 1, Post-emerging, a simple product is delivered from the niche idea to meet the emerging market. In Status 2, Post-diversifying, the product solution experiences diversification as central firms encourage more partners to contribute to its Solution platform, because the diversified ideas allow the ecosystem partners to interact in a high degree, which helps companies to find the best solution more efficiently and effectively in later phases. Diversified ideas also create a huge number of opportunities for different roles' participation. Every player, no matter in what position, can provide ideas and become involved in the new product development process.

In Status 3, Post-converging, various solutions that emerged in the co-evolution phase are selected for the specialised market. There is not only one best solution, but also some selected and typical solutions for industrial development. In Status 4, Post-consolidating, those selected products become dominant design products with well-accepted market standards in order to improve industrial efficiency and effectiveness. Finally, in Status 5, Post-renewing, new ideas are brought in to achieve the market transition.

In these six sequential statuses, the product types experience huge changes from ideas to dominant designs along these phases, and are adaptive to each phase's market requirements. As a result, 'Adaptive Solution' can be concluded as the key constructive element, which varies continuously along the BELC.

Solution platform

During the BELC, Solution platform has big impact only at the phase of Diversifying, Converging and Consolidating. In the phases of Emerging and Renewing, central firms focus on bringing niche ideas and their relevant products to renew the existing industry into the emerging industry. The other three phases require more collaboration among partners to nurture the new industry. So the Solution platform plays an important role in Diversifying, Converging and Consolidating phases.

In Status 2, Post-diversifying, the central firm has already introduced its Solution platform, which allows its partners to access the platform and deliver various product solutions very easily. In Status 3, Post-converging, many alliances are formed around the Solution platform in order to cope with the specialised market. As a result, partners are integrated around this Solution platform. Then, in Status 4, Post-converging, the Solution platform continues to improve and approach the dominant design so as to maintain the central firm's competitive advantage.

In conclusion, the Solution platform is one of the typical constructive elements that form the types of interaction among partners in the business ecosystem.

The platform has two levels of function. First, the Solution platform is proposed as a common component that is widely used in end-user products; for example, Nokia could adapt ARM's IP for mobile phones and Shanzhai OEM could adapt MTK's chip for mobile phones as well. Second, Solution platform provides an interface for other partners to work together for the future products.

By reviewing the main cases, it is believed that ARM, Intel and MTK are chip platform providers. Furthermore, Intel also manufactures chips. Regarding the core business of chip design, there are three typical types of solution platform. By using the project of LPS (1b) as an example, ARM co-designed its IP core with their leading partners, and then provided an interface to connect with other partners. This project indicates that ARM partly opened its Solution platform to its leading partners. In terms of the Atom project (2c), Intel closed the Atom chips platform but provided an interface to connect to its partners. However, they organised a supplementary software ecosystem to support the Atom platform. In summary, Intel closed its platform but organised a supplementary component as a platform to support its own products. Regarding the Turnkey model (3b), MTK developed further from ARM and Intel by integrating its chip core and basic software, and providing an interface to connect to other partners. Other projects are also reviewed and placed in Table 9.6.

Table 9.6   Solution platform classification

Types projects	Less open platform	Supplementary open	Turnkey model
1a			
1b			
1c			
2a			
2b			
2c			
3a			
3b			
3c			

Note: The blocks with colour mean each project has different types of platforms. Sometimes, one project's platform changes from one type to another type.

In terms of 2c, the Atom platform was initially closed. However, due to the complicated market requirements, Intel began to adopt its partners' feedback and embedded them into the new generation platform. As a result, project 2c was placed both in less open platform and supplementary open blocks.

MTK also began to collaborate with firms owning supplementary components *and* get feedback. For example, they began to contribute to the open source android platform *and* collaborate with carriers. So project 3c was placed into both supplementary open and Turnkey model.

By learning from those projects, three typical Solution platforms can be concluded as follows:

- Less open platform: central firms open their Solution platforms to a few key partners to co-develop the final products. They also provide support to the supplementary partners.
- Supplementary open: central firms close their Solution platforms, but with interfaces to connect to their partners. However, they initially coordinate with partners in order to provide supplementary components to their Solution platform.
- Turnkey model: central firms provide the closed platforms by integrating their core chips and basic supplementary components, but with an easy interface to connect to their partners.

Extended resources

This part describes how firms approach their extended complementary resources for delivering final products. In Table 9.7, resources are listed

Table 9.7 General constructive elements of business ecosystem

Constructs	Status 0: Pre-emerging	Status 1: Post-emerging	Status 2: Post-diversifying	Status 3: Post-converging	Status 4: Post-consolidating	Status 5: Post-renewing	Constructive elements
Structure	Niche idea	Emerging solution	Diversified solution	Selected solution	Product as dominant design	Niche idea:	**Adaptive Solution**
	Separated roles	Role (central firms with value chain initiated)	Roles (central roles with crossed value chain)	Roles (central firm with selected value chain)	Roles (central firm with integrated partners)	Roles (central role to reorganise partners)	**Functional Roles:** Initiator, Adopter, Specialist
			Platform with easy access	Integrated Solution platform	Continuing improvement of Solution platform		**Solution platform:** Less open; supplementary open; Turnkey model
		Extended existing resource	Various resource combination	Selected resources	Integrated resource		**Extended Resources:** partner network, financial resources, technology accumulation & industrial experience etc.
Infrastructure	New vision initiated	Common vision initiated	Vision co-designed	Finalised vision		new vision introduced	**New Vision Development**
		Leveraged partners	Diversified relationship with partners	Selected partners	Close relationship with partners		**Network governance**
		Platform-based simple value chain	Platform-based flexible network	Platform-based converging network	Close network around platform		**Core business process**
		Emerging market	Market diversity and flexible societal and policy system	Market specialised and adaptable societal and policy system	Mature policy and supportive societal system, stable market	Market transition	**Enabling mechanism development:** Market, societal and policy system etc.

only from Status 1 to Status 4. In the Pre-emerging and Post-renewing status, central firms are focusing on initiating new ideas to renew the existing market, so will not consider the complementary resources until they own new ideas to renew the existing market.

In Status 1, Post-emerging, central firms acquire existing resources to formulate a simple supply chain. Then, in Status 2, Post-diversifying, various resources are available and together with the supporting partners, diversified products can be proposed. In Status 3, Post-converging, as the market is specialised, various resources are selected and arranged by central firms and their partners for the typical products. Then in Status 4, Post-consolidating, as the dominant design is confirmed, various resources are integrated in order to trigger mass production. In these four sequential statuses, the resources are re-organised by partners in order to meet the firm's strategy along the BELC. As a result, 'Extended Resources' could be concluded to be the key constructive element, which is supporting the firm's development.

Besides the changing status, there are many different types of existing Extended Resources within the business ecosystem: 1) partner network: MTK's Shanzhai manufacturing network; Intel's PC network partners are also supporting resources to make them penetrate into the new market quickly and efficiently; 2) financial resources: Intel also offers financial support to its partners' network; 3) technology accumulation and industrial experience: as these cases indicate, industrial transition from mobile, PC industry into mobile computing industry. As a result, industrial experience and expertise are crucial for their future strategies.

9.4.2   Infrastructure side

The structural elements of business ecosystem are those stable factors that have a long-term orientation, while the infrastructure elements are the operational factors which may require decision-making on a day-to-day basis.

New vision development

In the business ecosystem context, firms not only focus on single product solutions, but also develop future visions of products to encourage contributions from partners. In Status 0, firms initiate the future vision for the development of the emerging industry. Then, in Status 1, Post-emerging, the central firm will share the vision with its potential value chain partners and persuade them to adapt to its Solution platform, while in Status 2, Post-diversifying, partners are also encouraged to co-design the future vision by bringing in diversified ideas. Different companies offer their own ideas to enrich the performance of final products as well as the ecosystem's vision. In Status 3, Post-converging, the industrial vision is finalised in order to select the potential best solution, which is accepted by most of the partners. Then the Consolidating phase achieves mass production of the product rather than dealing with the product vision. However, at the Renewing phase, Status 5, a niche vision is brought in that connects with a new Phase 1. Then, in the first phase, firms create visions and they are shared with one another.

In conclusion, by looking through the status of vision, 'new vision development' is the right phrase to demonstrate how firms encourage partners to contribute to future product development.

Network governance

At the Pre-emerging and Post-renewing point, firms develop niche ideas rather than initiate relationships with their partners. So in these two statuses, the relationships do not generate effects. Then, at the point of Status 1 as Post-emerging, the central firm has already leveraged in non-direct partners other than direct suppliers and customers in order to build up the supply chain for commercialising the new idea. At the point of Status 2, Post-diversified, the central company stimulates partner initiatives and encourages their contributions. The relationship between partners is diversified and flexible. At the point of Status 3, Post-converging, partners are selected around the central firms Solution platform for typical solutions. Finally, in Status 4, the relationship between the partners is integrated and stabilised for mass production.

By learning from the leveraged partners, diversified relationships with partners towards selected partners and integrated partners, 'Network governance', is used to describe those changes along the life-cycle phase.

Core business process

By expanding their relationships and sharing their visions in common with partners, central firms begin to consider that their core business process issues are to deliver their own products and form the collaboration among partners as well. In Status 1, Post-emerging, central firms' main ideas are to persuade partners to quickly commercialise their solutions. In Status 2, Post-diversifying, the core business process varies from company to company in order to meet the requirements from the market and the partners. A flexible network is adopted to deal with the different products and services around the central firm's Solution platform. This core business process is flexible and agile. In Status 3, Post-converging,

Table 9.8   Definition of constructive elements

	Constructive elements	Definition
Structure	**Adaptive solution**	The output of business ecosystem, which varied and adapted to the requirements of different phase of BE
	Functional roles	Players who act as different roles in Business Ecosystem (BE): initiator: those who are willing to build BE with their platform and product; specialist firm: those who are adding value to central firm's platform; adopter: those who build up final products by adopting initiator and specialist's co-designed platform
	Solution platform	Firms set up an interface in order to enable partners' involvement and encourage more partners' contribution
	Extended resources	Firms use existing and potential complementary resources to enable the innovation activity
Infrastructure	**New vision development**	The new image of industry future that central firm inspires in other partners
	Network governance	Regarding to BELC, the interaction between central firms and their partners evolves with different style
	Core business process	The key business and operation process to implement the new product development around Solution platform
	Enabling mechanism development	Several enabling factors help firm seek for business opportunities including policy, societal system and market trend

companies have already selected some typical solutions and then they prefer to form a converged network to increase industrial efficiency by changing from the diversified and flexible network. This kind of core business process is suitable when companies have clear objectives and future directions for the development of the industry. Then in Status 4, Post-consolidating, as the best solution is being identified, the core business process is finalised around the central firm's platform.

In summary, the core business process is the process relating to the central firm's new product development and commercialisation, which changes from the process of building a simple value chain, to enabling a platform based flexible network, then making the network converge, and finally towards forming a stable network collaboration appropriate to the sequential phases of the BELC.

Enabling mechanism development

Besides the constructive elements, there are also some external change factors, such as market status, policy and societal system which shape a suitable business environment and allow firms to collaborate with partners and capture business opportunities.

In terms of market trend, the market is always changing along the BELC. The market is emerging from existing relevant markets in Status 1, and then becomes diversified, specialised and stable in the following Status 2, Status 3 and Status 4. Then in Status 5, Post-renewing, the market experiences change in order to achieve market transition instead of recession. As a result, knowledge of market trends help firms to forecast market demand and to adopt appropriate strategy.

Policy and societal system build up the boundaries of business activities and offer great support as well. For example, in Status 2, Post-diversifying, policy and societal systems are flexible in order to nurture the emerging industry. Then in Status 3, as the market becomes specialised, policy and societal system become very adaptable to changes. Finally in Status 4, Post-consolidating, as the market became stable, policy would be highly mature and societal system would also be very supportive.

In conclusion, by reviewing these factors of market trend, policy and societal system, the term 'Enabling mechanism development' could be used to describe their adaptation to the sequential phases of BELC.

All of these elements from the structure and infrastructure side are finally generated as the fundamental factors of a business ecosystem. From the discussion above, the definitions of each element are identified in Table 9.8.

9.4.3   Integration of constructive elements

Those eight elements have internal relationships to compose a business ecosystem as shown in Figure 9.1. The element of 'new vision development' is the driving force of an ecosystem evolution. The business ecosystem was very complicated as its nature is composed of different levels of organisations. Thus it is hard to persuade all the partners move forward in the same direction with formal authorities such as the contract, alliance or other agreements (Gulati et al. 2012; Reid & Brentani 2012). Instead it is possible to move those partners with informal authorities such as the vision of a business ecosystem. As a result, it is better to encourage the partners by sharing the vision of the future and driving them together along the way.

Normally the focal firm has two ways to transfer the vision into the adaptive solution. The upflow of Figure 9.1 demonstrates how the focal firm provides the way of 'How' to work with partners. Thus they propose the core business process and co-design the platform with partners. The down flow of Figure 9.1 shows with whom focal firm could work. Hence, the focal firms coordinate the different partners with specialised functions and finally generate a network. Normally the upflow and downflow have interacted as well: the core business process proposed by focal firms will specialise those functional roles' position. However, the functional roles might be less capable of impacting on the core business process; the solution platform will formulate the partner network and govern the network.

Then the output of a business ecosystem is not only an adaptive solution but also the 'extended resource' which could recontribute to the further business ecosystem evolution.

By coping with the emerging demand and challenges, the typical enabling mechanisms such as the mechanisms of policies and societal systems will enable ecosystem partners to develop new vision to tackle the challenges and demand.

9.5   The natural metaphor

The term 'Business ecosystem' was first introduced by Moore in 1993 as a metaphor of the mechanisms of a natural ecosystem (Moore 1993). However, Moore did not show the mapping relationship between

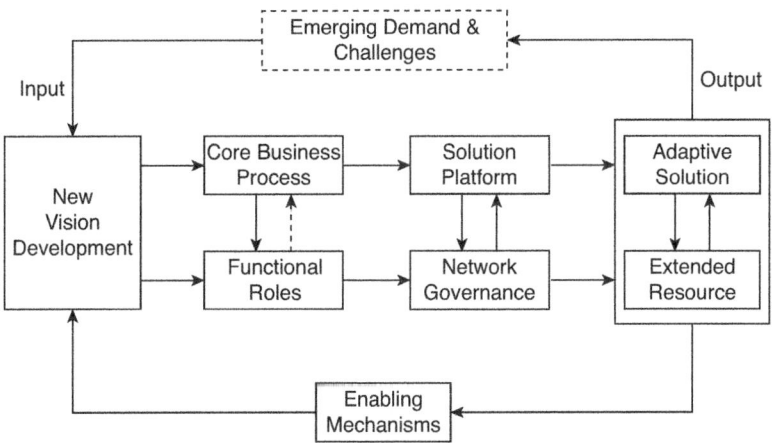

Figure 9.1   Integration of business ecosystem constructive elements

business ecosystem and natural ecosystem. This section lists a preliminary mapping framework which aims to enrich the ecosystem metaphor on a detailed level from the constructs perspective.

From Table 9.9, the elements are listed both in business ecosystem and natural ecosystem (Odum 1971). Roles of the organisms in a natural

Table 9.9 Comparison between business ecosystem and natural ecosystem

	Business Ecosystem (BE)	Relationship between BE and NE	Natural Ecosystem (NE)
Structure	Adaptive Solution	Solution developed by the supply chain partners like energy and material along food chain	Material and Energy like Carbon, Nitrogen
	Functional Roles	Different roles interact within ecosystem	Autotroph (Producer), heterotrophs (Consumers) etc.
	Solution platform	Central firm initiates business ecosystem by its Solution platform which demonstrates ecosystem scope and place	Ecosystem scope and place
	Extended Resources	There are also many resources to support activities in business and natural world	Water, air, etc.
Infrastructure	New Vision Development	Vision and symbiosis demonstrate the co-evolution process between roles	Symbiosis
	Network governance	Relationship between roles in business and natural world demonstrates selection process	Selection
	Core business process	Core business process and energy and material cycle both demonstrate how ecosystem develops its product or energy and materials.	Energy and material cycle
	Enabling mechanism development	There are also many other factors that have impact on ecosystem development	Non-living environment and occasions, etc.

ecosystem like producer and consumer, predator and prey are just like those in business ecosystem. In a natural ecosystem, there are energy and materials along the food chain which are similar to the product solution adapted by different tiers of partners to form a value chain. The area of interaction among organisms in a natural ecosystem reflects the platform concept in business ecosystem. Furthermore, other materials like air and water could be the complementary resources for natural ecosystem's development.

On the infrastructure side, selection processes exist in a natural ecosystem which matches the situation of network governance in a business ecosystem. Network governance reflects all the interaction and selection among partners. Symbiosis phenomena can easily be found; one example would be the relationship between the clownfish and the sea anemone. The sea anemone protects the clownfish while the clownfish feeds the anemone by gathering nutrients and leaving nutritional waste. Companies work together under a common vision through the life cycle which demonstrates the symbiosis mechanism. The core business process shows how a business ecosystem develops its products, which matches the energy and material cycle in a natural ecosystem. The last factor is the non-living environment and the scope for natural organisms to adapt their surroundings against uncertainty, which parallels the 'enabling mechanism development' of altering policy, societal issues and market trends as well.

In summary, the metaphor study here aims to inform the understanding of business ecosystems by learning from the understanding of natural ecosystems, especially from the construct perspective. Those constructive elements have a big impact on ecosystem development both in the natural and business worlds. Comparison is an aid to understanding the comprehensiveness of business ecosystem constructs and the internal logical relationship between its elements.

Besides the constructs perspective, Iansiti and Levien also made a comparison between natural and business ecosystems. They highlighted three criteria to measure the health of an ecosystem: robustness, productivity and niche creation, which exist in both natural and business ecosystems. Moreover, they highlighted differences: a business ecosystem has a purpose of doing business, while a natural ecosystem does not have a purpose of doing anything (Iansiti & Levien 2004b).

9.6 Discussion on constructs study

Previous literature suggested that the nurturing process or strategy along the life cycle were influenced by constructive elements, yet there were no sufficient studies on constructing a business ecosystem. Even though Moore proposed seven strategies and eight campaigns, those works did not focus on the constructs level, but on activities level.

In this study, there are two distinct points to highlight in the constructs study at the business ecosystem level: first, constructive elements are studied in the context of life cycle, which aims to clarify the key constructive elements that vary in each phase with different impacts. Secondly, It also maintains a close relationship with the following theories, and extends their models of structure and infrastructure from the firm level (Hayes & Wheelwright 1984), intra-firm level (Shi & Gregory 1998), inter-firm level (Harland 1996) and internationalised inter-firm level (Zhang et al. 2007; Srai & Gregory 2008) towards the business ecosystem level as in Table 9.10. In detail, there are four aspects to highlight as contributions in comparing this business ecosystem construct study with previous literature.

- **Functional roles and their relationship:** constructive elements at the business ecosystem level present the different organisations' role types (initiator, adopter and specialist) and their interaction along the life cycle. The types of roles can help firms easily understand their functional types and position in order to implement appropriate strategies. It also presents the different relationship structures at the network level (which demonstrates the wide scope of relationships among different levels of organisations).
- **Future vision and industry solution:** the business ecosystem highlights new vision development which focuses more on developing future direction with partners than on the current new product development level. Also the business ecosystem regards an adaptive solution as a constructive element, although it is normally viewed as the output of a system rather than as a constructive element itself. This is because an adaptive solution in the business ecosystem demonstrates a co-development process among different levels of organisations, with the purpose of meeting different market environments along the life cycle.
- **Solution platform and Core business process:** A Solution platform forms the pattern of partners' interactions and enables solution diversity regardless of geographic concerns. All of the Core business process in the business ecosystem is also implemented with reference to the Solution platform.
- **Extended Resource and Enabling mechanism development:** different from previous manufacturing systems and partner networks,

Table 9.10 Construct study evolution

	Firm (Hayes & Wheelwright 1984)	Intra-firm (Shi & Gregory 1998)	Inter-firm (Harland 1996)	International inter-firm level (Zhang et al. 2007)	Business ecosystem
Structural	Capacity, Facilities and Technology	Factory as a node of network	SN Actors configuration	Network structure	Functional roles
		Geographic dispersion	Capacity	Coordination mechanism	Solution platform
		Horizontal coordination			
	Vertical Integration	Vertical integration	SN facilities configuration		Extended Resource Adaptive Solution
			Do or buy		
Infrastructural	Organisation	Organisation/ evolution	Network organisation	Governance system	Network governance
	Production planning/ material control	Operation /control	SN operations planning and control		Core business process
	Quality	Learning/knowledge sharing	SN quality systems	Support system	
	Workforce	Response / exploration	SN human resource policies		
			Performance measurement		
			New product/ service development		New vision product
					Enabling mechanism development

the business ecosystem highlights the importance of exploring complementary resources and the unique concerns about Enabling mechanism development, including market status, policy environment and societal system. Those factors also have big impacts on the new industry development.

9.7　Conclusion

This chapter developed the theory of constructing a business ecosystem. General constructive elements were identified to enrich business ecosystem theory by using a classical framework of 'structure–infrastructure'.

First, nine sub-cases were analysed in the context of the six phase statuses along the BELC. Regarding the difference of each phase status, several constructive elements were highlighted in each phase.

Secondly, by comparing the constructive elements of the six phase statuses, the elements were generalised in order to formulate the business ecosystem as a whole. On the structure side, Adaptive solutions, Functional roles, Solution platforms and Extended resources were identified, while New vision development, Network governance, Core business process and Enabling mechanism development were discovered on the infrastructure side. Those eight constructive elements integrate to compose a business ecosystem.

Thirdly, the ecosystem metaphor was developed by comparing business ecosystems with natural ecosystems in order to gain a more comprehensive view of ecosystem development and its mechanisms.

Finally, a discussion was conducted by comparing the constructs study from the levels of firm, intra-firm, inter-firm and internationalised inter-firm towards business ecosystem. Four points were listed as the contributions.

10
Business Ecosystem Configuration Pattern

10.1 Introduction

This chapter categorises the different configuration patterns of business ecosystems through the data gathered from exploratory and main cases. The outcome of this chapter focuses on the following objectives:

- To identify business ecosystem configuration patterns by two dimensions reflected from two of the key representative constructive elements – Solution Platform Openness and Solution Diversity.
- To picture the evolutionary path of each main case on nurturing business ecosystems with their selected patterns.

The configuration pattern study has been mentioned by a few papers, but has not been well explored and highlighted in previous studies (Shi & Gregory 1998; Zhang et al. 2007). There are two reasons to suggest that each business ecosystem has its own configuration pattern.

First, configuration patterns provide a good way of concisely representing a complex organisation and forming an integrated view which can be identified by the constructive elements of the 'structure and infrastructure' framework (Shi & Gregory 1998). Such thinking was adopted by Zhang's framework (Zhang et al. 2007) which used network structure, coordination mechanisms, governance systems and support systems to categorise a GEN (global engineering network). To date, configuration patterns have been studied in firm-based manufacturing systems, international manufacturing networks, supply chains and GENs.

Second, three of the main cases experienced different results after passing through the five sequential phases of the business ecosystem life cycle (BELC). The stories of success or failure can demonstrate different firms' strategies as they evolve along the BELC. Additionally, configuration pattern has much impact on firms' strategies (Hayes & Wheelwright 1984; Shi & Gregory 1998; Shi & Gregory 2001). Therefore, the study of configuration pattern is useful as a synthesised way to present the typical strategy of a firm's nurturing of a business ecosystem.

From a practical perspective, it is useful to identify configuration patterns, as business ecosystems are complex and difficult to comprehend. With the help of typical pattern identification, both industry and academia can capture the key features of a specific business ecosystem. Hence, it is worthwhile to study the configuration patterns of business ecosystems in order to understand a firm's strategy for nurturing their business ecosystem from an uncomplicated and integrated view.

Appendix Figures 1, 2, 3 list all the activities of the three main cases along the BELC. Each company follows their own strategies to lead activities, which form the different patterns as these activities reflect the constructive elements of a business ecosystem. This chapter follows the strategies by using the dimensions from constructive elements to categorise different patterns, as below. Hence, the configuration pattern of a business ecosystem can be identified by the following steps:

- First, select the key and most representative dimensions to identify a configuration pattern: these two dimensions, which may include Solution platform and Solution Diversity, are used to categorise different patterns. The dimensions reflect industrial requirements such as interoperability and uncertainty respectively.
- Second, devise scales along which to measure the two selected dimensions. Differences between cases' configuration patterns can be shown by placing the cases appropriately along the scales. The patterns will reflect the differences between each ecosystem's strategy.
- Third, to study pattern evolution along the BELC: this section will include an in-depth study of the three main cases' evolution over pattern selection and demonstrate different pathways of nurturing ecosystems.

10.2   Dimensions for identifying ecosystem pattern

From the industrial background chapter, two dimensions are highlighted as the key drivers for business ecosystem thinking as illustrated in Table 10.1. The first is about uncertainty and includes market, application and technology perspectives. The second is about interoperability and includes platform and supplementary interoperability.

Table 10.1 Relationship among drivers, requirements and constructive elements

Industrial drivers	Items	Requirements ➡	Categorising ⬅ dimensions	Constructive ⬅ elements
Uncertainty	Market uncertainty	Adopt and initiate market requirement	Solution Diversity (end-user product)	Adaptive Solution
	Application uncertainty	More choice on ideas, diversity and openness		
	Technology uncertainty	Central companies with different architecture or platform		
Interoperability	Platform interoperability	More acceptable architecture or platform	Solution Platform Openness	Solution platform
	Supplementary interoperability	Easy interaction around platform; platform openness		

In terms of uncertainty issues, companies enable solution (end-user product) diversity as a priority to meet the various and uncertain requirements. Regarding market uncertainty, companies can acquire useful information or initiate a market requirement. Diversified ideas or solutions are encouraged to deal with application uncertainty. As a result, central companies can provide the technology platform or architecture and work together with partners to reduce the uncertainty of technology. So Solution Diversity is used to describe the output that a business ecosystem generates in order to cope with uncertainty issues.

Interoperability aims to describe the need for companies' interaction in order to cope with uncertainty issues. Platform interoperability tells the story about the central firm's strategy on platform building. Acceptance of this platform improves the industry's diversity and efficiency. Regarding supplementary interoperability, non-central firms' should be more easily able to interact around the platform and they can also contribute to that platform. This dimension focuses on platform building and its interactions, which is linked to the constructive element Solution Platform Openness. Solution Platform Openness also reflects how firms organise the partners' network in a business ecosystem.

Table 10.1 describes the relationship between drivers, requirements and the two categorising dimensions (Solution Platform Openness and Solution Diversity). Solution Platform Openness reflects the degree of partners' interoperability inside the business ecosystem context. Solution Diversity deals with three kinds of uncertainty along the life cycle. Meanwhile, 'Adaptive Solution' and 'Solution platform' are two key constructive elements of business ecosystems, and demonstrate the key features of business ecosystem operation.

Solution Diversity and Solution Platform Openness are factors fundamental to the categorisation of configuration patterns of business ecosystems.

10.3   Configuration pattern identification

10.3.1   Cases database

Appendix Figures 1, 2, 3 illustrate the main cases with different strategies at sequential phases of BELC. Each project has its different configuration pattern at different phases of the life cycle. Following this idea, these are 10 sub-cases by phase in ARM, 12 sub-cases by phase in Intel and 12 sub-cases by phase in MTK. Thus, in total there are 41 cases, including the 7 exploratory cases.

These cases are named both by project name and phases' position. For example, in ARM's leader partners strategy project, they are 1b(2), 1b(3) and 1b(4), while in MTK's smartphone project, they are 3c(2) and 3c(3).

10.3.2   Cross-cases analysis

The projects and seven exploratory cases (projects) form the case database to demonstrate the sets of configuration pattern. Table 10.2 lists the 41 cases, where the main cases are arranged by phases and the exploratory cases by general characteristics. These projects will be categorised by the two dimensions of Solution Platform Openness and Solution Diversity.

By cross-cases analysis, Solution Platform Openness and Solution Diversity are described with different measures. For example, case 23 (MTK VCD) uses a very closed platform and single solution, while in the case of 34 (MTK Smartphone), the platform is less open and the solution has fewer features, some of which can be altered by the partners.

In case 35 (Linux), the open community ecosystem and its platform is open to all volunteers and companies with various solutions. Hence, in this case the platform is fully open and the solution has many features. However, in some other cases, the degree of openness and diversity do not match well. For example, case 40's (Apple online store) platform is fully self-designed although with an easy access interface. Partners cannot change the platform's structure, but can build application software based on the platform. As a result, the solution is of high variety although the platform is closed. Through the observation of these cases, platform openness is divided into three levels: 'close', 'less open' and 'open', while Solution variety is divided into three levels: 'scarcity', 'middle' and 'abundant'.

In Solution Platform Openness, the score 'open' means that the Solution platform can be co-designed by every participant and the score of 'less open' implies it can be designed by selected partners while the score 'close' means that Solution platform is only designed by the central firm. In Solution Diversity, the score 'abundance' means that end-user products vary, the score 'middle' means end-user products are controlled by several different solutions, while the score 'scarcity' means that end-user products are dominated by a very limited solution.

Hence, the 41 cases can be placed into a matrix as depicted in Figure 10.1. Seven types of configuration pattern (from a possible nine) evolved from the 41 cases. (Cases in each block appear at the same level because of the two scoring dimensions). For example, number 2, 6, 9, 21, 33 and 38 are in the same block and hence demonstrate the ecosystem of a 'less open' Solution platform and 'abundant' solutions).

10.3.3 Configuration patterns identification

In Figure 10.1, the 41 cases are positioned in seven blocks. Those seven blocks are named following the numbers as Pattern 1 to Pattern 7 (bottom row, left to right – patterns 1, 2, 3; middle row, centre to right – 4, 5; top row, centre to right – 6, 7). Cases cluster at Patterns 2, 3, 4 and 5.

Previous industry developments like the PC and VCD industry are not very complicated. Firms intend to approach dominant designs and allow partners to develop solutions based on the dominant designs. Seventeen out of 41 cases are positioned at Pattern 2 and Pattern 3, where the central firm's Solution platform is closed but the solutions based on that platform score from 'less abundant' to 'abundant'.

As the mobile computing industry becomes more complicated, compared with PC and VCD industry, central companies gradually open

Table 10.2   Configuration pattern classification

Case No.	Sub-case (life cycle phase)	General characteristics	Solution platform (Openness)	Solution (Diversity)
1.	1a(1) mobile	– ARM7 introduced – leverage partners for mobile	Less open	Middle
2.	1a(2) mobile	– Leader partner with TI – aim to involve more partner	Less open	Abundance
3.	1a(3) mobile	– Strong alliance with TI and Nokia – Application win the advantage	Less open	Middle
4.	1a(4) mobile	– Continue to improve IP – Mass production for mobile	Less open	Middle
5.	1a(5) mobile	– ARM7 redesigned to Cortex – Find new partners	Less open	Middle
6.	1b(2) LPS	– Co-design on IP architecture – Co-marketing – Connected community	Open to leader partners (less open)	Abundance
7.	1b(3) LPS	– Luminary platform – Co-marketing – Succeed from ARM7 software ecosystem	Less open	Middle
8.	1b(4) LPS	– More leader partners – More than 100 customers	Less open	Middle
9.	1c(3) IP categorised	– Customised into three IP groups – Market specialised	Open to leader partners (Less open)	Middle
10.	1c(4) IP categorised	– IPs continue improving – Partners re-organised to three streams	Open to leader partners (Less open)	Middle
11.	2a(1) PC	– Focus on PC processor – key supplier to IBM PC	Close	Middle
12.	2a(2) PC	– Intel support different OEM's PC – Support non-direct customers	Less open	Abundant
13.	2a(3) PC	– PCI platform strategy – Encourage partners to adopt, compatible to different generation	Less open	Middle
14.	2a(4) PC	– Continue to introduce platform – Mass production – Close alliance with Microsoft	Less open	Middle

Continued

Table 10.2 Continued

Case No.	Sub-case (life cycle phase)	General characteristics	Solution platform (Openness)	Solution (Diversity)
15.	2a(5) PC	– StrongARM IP – Encourage partners	Close	Middle
16.	2b(1) Xscale	– Used ARM's architecture – Encourage PC partners	Close	Scarcity
17.	2b(3) Xscale	– Collaborating with Motorola – Finalised design	Close	Middle
18.	2b(4) Xscale	– Outsource manufacturing to TSMC – Fast and stable supply chain	Close	Middle
19.	2b(5) Xscale	– High maintenance cost – Big competitors	Close	Middle
20.	2c(1) Atom	– Incubating ideas; Atom chips – Succeed from PC industry	Close	Middle
21.	2c(2) Atom	– Initiate concept of MID – Support partner and Moblin ecosystem	Less open	Abundance
22.	2c (3) Atom	– Introduce Netbook standards – Select partners to follow – Keep refining Atom	Less open	Middle
23.	3a(1) VCD/DVD	– Single chip solution – Technology from its background	Close	Scarcity
24.	3a(2) VCD/DVD	– Single chip solution – Partners from Hshinchu park	Close	Abundance
25.	3a(3) VCD/DVD	– Single chip solution – Work with VCD and PC players in mainland China	Close	Middle
26.	3a(4) VCD/DVD	– Improving solution – Downstream supply-chain innovation	Close	Middle
27.	3a(5) VCD/DVD	– Use DVD to replace VCD – Fast succeed previous network	Close	Middle
28.	3b(1) Mobile	– Turnkey solution – Taiwan local OEM	Close	Scarcity
29.	3b(2) Mobile	– Turnkey solution – Enabled Shanzhai network innovation	Close	Abundance
30.	3b(3) Mobile	– Turnkey solution – Support local IDH and provide training	Close	Middle

Continued

Table 10.2   Continued

Case No.	Sub-case (life cycle phase)	General characteristics	Solution platform (Openness)	Solution (Diversity)
31.	3b(4) Mobile	– Continue to improve Turnkey with more function – 20% China shipment	Close	Middle
32.	3b(5) Mobile	– Introduce Smartphone solution – Succeed the Shanzhai network	Close	Middle
33.	3c(2) Smartphone	– Product design together both software and hardware – Join Android – Connect with carrier	Less open	Abundance
34.	3c(3) Smartphone	– Improved Turnkey model – Selected partners – Support and training	Less open	Middle
35.	Linux	– Open community – Volunteer contribution – Or companies contribution	Open	Abundance
36.	Symbian	– Open community – Coordinated by few companies	Open	Middle
37.	TSMC	– Design for manufacturing ecosystem – Collaborate with EDA and IP companies	Less open	Middle
38.	China Mobile	– OMS operating system ecosystem – Encourage companies' contribution	Less open	Abundance
39.	VIA GMB	– Integrated platform – Integrated supply chain for Netbook	Close	Abundance
40.	Apple online	– Close platform and open interface – Online contribution from different companies	Close (self-design)	Abundance
41.	Marvell	– Integrated platform – Integrated supply chain for final device	Close	Abundance

their Solution platforms to attract more partners to develop applications. As a result, nearly half the cases cluster in Pattern 4 and Pattern 5, where the Solution platform is less open and the solutions vary from 'less abundant' to 'abundant'.

There are three cases clustered at Pattern 1. These three demonstrate how firms penetrate into a mature industry with closed platform and scarce solutions. Case 16 shows how Intel used ARM's IP by copying its dominant strategy in the PC industry while case 23 and 28 tells the story of MTK's Turnkey solution.

Additionally, there are the boundaries between Pattern 5 and Pattern 6. Under Pattern 5, the configuration pattern is more business oriented while the upper part of Pattern 6 is more volunteer oriented. At times, companies or volunteers cross boundaries for particular purposes. Intel used the Linux foundation (Pattern 7) as the base for the Moblin ecosystem to support Atom chips with various operating systems. The Symbian foundation was built based on the business – Symbian software. The purpose is clear for these two sides: volunteer orientation is to enrich the base while business orientation is to make good use of this base. They can benefit from each other.

Besides these seven patterns, in Figure 10.1, there are two empty blocks with no cases. There is some paradox in these two blocks in the

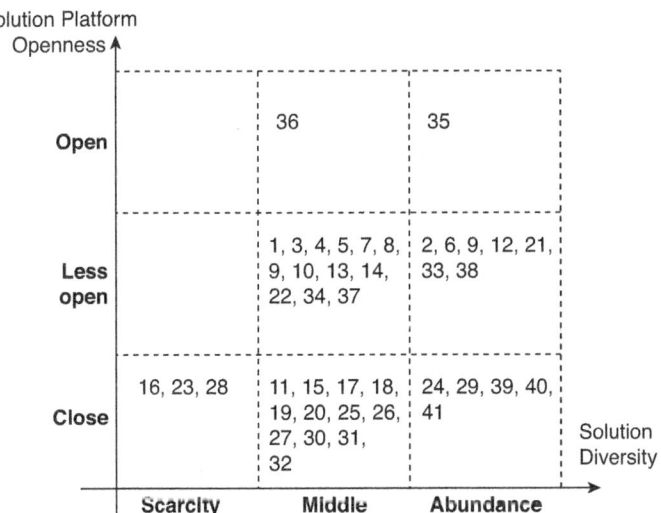

Figure 10.1   Configuration pattern classification

business world. If the platform is less open or open, it means several partners are involved in co-developing the new platform. As a result, the solution based on the platform would be changed and diversified as the partners wish, especially at the open platform level, where there may be several solutions from different volunteers and companies. So the pattern with the product that is less open or open but solution is limited does not exist.

A detailed description of the seven patterns is listed in Figure 10.2.

10.3.4 Seven configuration patterns strategies

From the two dimensions of Solution Diversity and Solution Platform Openness, two issues echo through the learning from the real cases.

Innovation scope: solution diversity

The levels of 'middle' and 'abundance' demonstrate the degree of Solution Diversity. On the 'middle' level, companies are selected to develop several products based on the key platform provided by the central firm. As a result, the solution has less abundance. However, on the 'abundance' level, companies are welcome to develop new products based on the central firm's Solution Platform, which can cut the entry barriers and improve industrial efficiency, making the network very flexible. Hence, the innovation scope is expanded for partners already in the network. In conclusion, Solution Diversity tells the narrative about how a central firm can expand the innovation scope and involve more practitioners.

Power distance: platform openness

The level of platform openness sets the entry barrier to a platform's development. In a less open platform, it is controlled by only a few central companies, so the development strategy is under the scope of their intention. Partners do not have bargaining power on the platform development level. In the open platform, the story is different. All partners are encouraged to enrich the platform with no entry barrier. The bargaining power vanishes in order to attract more volunteers and business partners. So in this dimension, platform openness mostly focuses on the power distance or entry barrier among different level partners in the business ecosystem.

In conclusion, there are seven types of business ecosystem configuration categorised by Solution Diversity (innovation scope) and platform openness (power distance). They are: simple solution ecosystem, platform enabling ecosystem, platform integrated ecosystem, platform

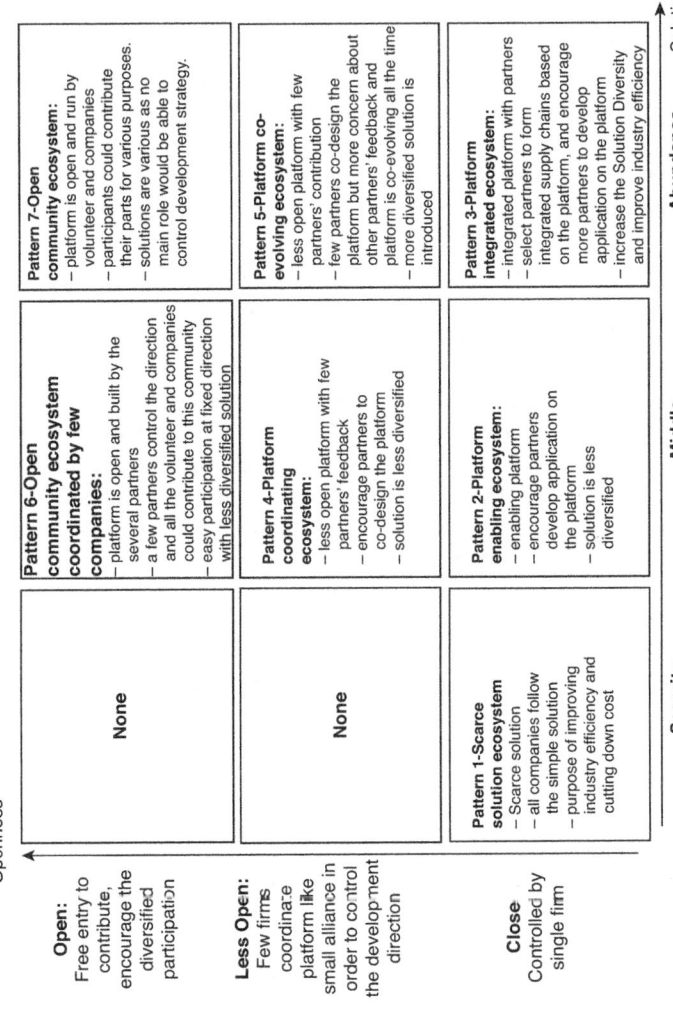

Figure 10.2 Conclusion on business ecosystem configuration pattern

coordinating ecosystem, platform co-evolving ecosystem, open community ecosystem coordinated by a few companies, and open community ecosystem.

10.4   Configuration pattern evolution along business ecosystem life cycle

In the last section, seven configuration patterns are identified as the type descriptors of existing ecosystems. From a dynamic view, this research also explores how those companies adopt a pattern type along the BELC. In this section, the evolutionary view of pattern selection will be identified based on a comparison of the main cases' ecosystem evolution. The amalgamation of Table 10.3 and Figure 10.2, Table 10.3 demonstrates three main cases' configuration pattern evolution along the BELC.

10.4.1   ARM pattern: openness as its nature

Generally, in ARM's ecosystem, ARM took Pattern 4 in Phase 1, acted as Pattern 5 in the second phase and Pattern 4 in the last three phases. ARM was a small firm in the early 1990s. In order to get support from big IC design firms, they provided a less open platform and persuaded partners to adopt their IPs. In the second phase, ARM used leader partners' strategy to encourage partners to contribute to the new ecosystem. So Pattern 5 was adopted as a platform co-evolving ecosystem. This pattern aims to deliver Solution Diversity and sustain the ecosystem development. In the third and fourth phase, IC design firms helped ARM to approach the best solution for industry development. However, regarding ARM's strategies, they took Pattern 4 to make their business ecosystem stronger

Table 10.3   Configuration pattern evolution from main cases perspective

Cases	Projects	Emerging	Diversifying	Converging	Consolidating	Renewing
ARM	1a	4	5	4	4	4
	1b		5	4	4	
	1c			4	4	
Intel	2a	2	5	4	4	2
	2b	1		2	2	2
	2c	2	5	4		
MTK	3a	1	3	2	2	2
	3b	1	3	2	2	2
	3c		5	4		
	Conclusion	Three 1; Two 2; One 4	Five 5; Two 3	Six 4; Three 2	Four 4; Three 2	One 4; Four 2

and robust. In the fifth phase, Renewing, ARM began to enter embedded market by upgrading ARM7 into Cortex-M. So in order to make good use of the former mobile ecosystem, ARM kept using Pattern 4 to encourage partners' contribution. ARM used the same pattern strategy in two other projects.

In conclusion, ARM's ecosystem shows a picture of value co-creation with partners. In each phase, ARM aimed at working with different partners. This was not only because ARM's products were far from the end-user products, but also because ARM aimed to enable partners to work around their platform and deliver various end-user products.

10.4.2   Intel pattern: gradually opening up

Intel was successful in the PC project, but not in the Xscale project, while the Atom project is still ongoing. Intel used Patterns 2-5-4-4-2 along five phases of the BELC in the PC industry. Intel introduced the PCI standard in Phase 3 to end Solution Diversity and approach the typical design. This was a successful strategy, which later allowed Intel to dominate the PC industry. Intel experienced the success as the dominating strategy in the PC industry. As a result, they simply copied this dominant strategy when they tried to penetrate the mobile phone market (Xscale project). They ignored the co-evolution process with partners, and introduced the industry standard as they had done in the PC industry. As a result, due to the lack of supporting partners, Xscale was not successful.

By learning from their previous experiences of success and failure, Intel realised the importance of the business ecosystem. Hence, they started to build a business ecosystem in the Atom project. Intel introduced the Netbook specification standard to dominate Netbook market. These standards were co-designed with several partners and enabled many other partners to accept the standards. Then Intel changed the strategy to re-enter the mobile market. In the mobile industry, more partners were involved and the solution was not finalised. Though Intel took Pattern 2 to start with, they then followed a similar strategy to ARM: Intel used Pattern 5 to increase the Solution Diversity and encourage participation.

Intel thus began to open its door to partners rather than dominating industry development. They found that they were not as strong in the mobile industry as in the PC industry. So in the second phase of the mobile computing ecosystem, they hoped to co-evolve with partners, as they aimed to rapidly set up the ecosystem and get as much partners' contribution as possible.

10.4.3   MTK Pattern: opportunity driven

MTK penetrated the new market when the market had already reached its maturity. So they often used Pattern 1 with Turnkey solution to attract a limited number of partners. By using Pattern 3, they penetrated into mainland China and enabled thousands of companies to work together, which also triggered the emergence of diversified solutions. In the following three phases, they selected partners and provided strong support to them in order to maintain their competitive advantage by using Pattern 2. Furthermore, as they dominated the current market, they could also penetrate into other relevant industries with similar strategies, even if the industry was not mature (DVD). As a result, from the VCD, DVD and mobile phone markets, MTK used strategies similar to their single-chip solution to improve industrial efficiency by using the pattern evolution model of 1-3-2-2-2.

However, as the mobile computing product became more complicated with various requirements, MTK began to get feedback and work with local partners as well. MTK just provided a less open platform and encouraged more partners' contribution on the platform as Pattern 5 and then Pattern 4 in phase three. Increasing openness was also the trend for other cases like ARM and Intel in the mobile computing industry.

From the general view of the MTK ecosystem, 'opportunity driven' could represent their development strategy very well. Due to the limited resources they had in different industries, they did not penetrate into the industry until it had matured with high volume shipment. MTK at this point made good use of its advantages on cost saving and technology integration to overcome competitors by providing the Turnkey solution with a low entry barrier. Additionally, at times, MTK was very responsive to local requirements by adding or changing some functions of its integrated solution.

Figure 10.3 compares each case's pattern evolution along the five phases of the business ecosystem. ARM adopts a very open pattern, Intel becomes open, while MTK has a very opportunity driven model. However, due to the data scope limitation, only three pattern evolution models are identified. For future research, more pattern evolution models can be found based on the study of more cases.

10.5   General path of ecosystem pattern evolution

This section aims to develop the general path of pattern evolution model by comparing the paths of the main cases' in Figure 10.3. Through the observation of each of the main cases illustrated in Table 10.3, the typical

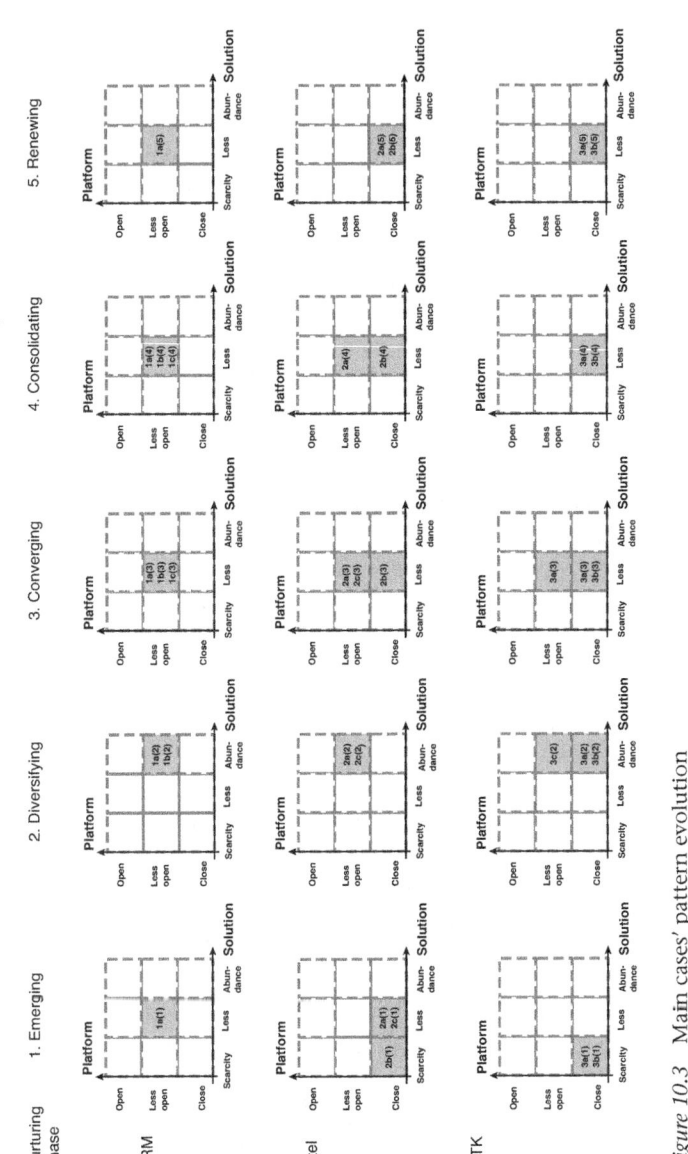

Figure 10.3 Main cases' pattern evolution

pattern selection path demonstrates the different strategies along the nurturing process. In summary there are two paths of pattern evolution in different contexts.

10.5.1   Context one – Industry Maturity: Pattern (1-3-2-2-2)

There are three projects with mature industry background: 2b (Xscale), 3a (VCD) and 3b (Mobile 2G). Even those projects implemented by different companies like Intel and MTK selected similar patterns along the five phases. As the mobile and VCD market matured, both introduced closed platform (Pattern 1) total solutions with the purpose of approaching design dominance. MTK was successful while Intel failed because at the second phase, Diversifying, it was very important to attract more partners to enable the products' variety, which would then receive strong support from partners in the proceeding phases. MTK endeavoured to enable the downstream partners' innovations and end-user product variety (Pattern 3) while Intel ignored this phase and directly entered into the phase of Converging with the aim of approaching design dominance as it had done in the PC industry. As a result, Intel failed and MTK succeeded.

As the industry was mature, in the following two phases, the key point was to push their Solution platform into design dominance and attract contribution from more partners. So they began to select partners with good performance and to deliver relevant products (Pattern 2). At the phase of Consolidating, central firms maintain the competitive advantage. MTK as a latecomer in the mature mobile market used the Turnkey model Pattern 2 to overcome the competition. At the final phase of Renewing, most companies were encouraged to renew the industry by bringing in ideas. MTK brought in the DVD as a new product to renew the VCD market in order to maintain their competitive advantage. So in that phase, companies also place themselves in Pattern 2.

10.5.2   Context two – Industry Emergence: Pattern (4/2-5-4-4-4/2)

Besides the last three projects, companies implemented the other six projects when the industries were uncertain or at early stages: 1a- emerging mobile phone market for ARM in the early 1990s, 1b- embedded market, 1c- new three streams market, 2a- new PC industry for Intel in the early 1980s, 2c- emerging mobile computing industry and 3c-emerging mobile computing industry (smartphone industry). Regarding the different main cases, they have different strategies in phase one and five, but the same strategies in phases two, three and four.

In the Emerging phase, the key purpose is to rapidly commercialise the emerging solution, so normally firms like Intel adopt Pattern 2 (Platform enabling ecosystem) as their first priority. Pattern 2 provides an enabling platform and also provides an opportunity for other companies to contribute in order to speed up the commercialisation process. However, as many industries become complicated and uncertain at the design stage – as in the embedded and mobile computing industry – firms like ARM prefer to adopt Pattern 4 (Platform coordinating ecosystem) to allow partners' participation in co-designing the Solution platform.

In phase 2 – Diversifying, most of the sub-cases are positioned in Pattern 5: platform co-evolving ecosystem. Pattern 5 provides a less open platform co-designed with the contribution of a few partners. Simultaneously, the central firms are also concerned about the other partners' feedback and allow the platform to co-evolve with the main partners and enable a diversified solution.

In the third phase – Converging, most companies place themselves in Pattern 4 to nurture the industry into maturity. As they experience the diversified platform co-evolving process, the platform becomes converged in order to approach the best solution. In this phase, Intel introduced industrial standards and ARM allied with leader partners.

In the fourth phase – Consolidating, firms aim to maintain the competitive advantage. ARM formulated strong alliances with leader partners while Intel introduced more industrial standards. They both used Pattern 4 to encourage their close partners to help maintain their competitive advantage.

In the final phase – Renewing, most companies are encouraged to renew the industry by bringing in a substitute. So in this phase, Intel aimed to expand its strategic dominance in the new market and they regularly adopted Pattern 2 to maintain their advantage. This idea will be co-developed in Phase 1 and Phase 2. Regarding the ARM case, they upgraded ARM7 into Cortex-M. They aimed at using leader partners' strategy later as the embedded market is very uncertain. As a result, they took Pattern 4 to renew their ecosystem.

From the above analysis, a generic picture can be drawn to demonstrate the general strategy of the configuration pattern evolution along the growth of the business ecosystem as illustrated in Figure 10.4.

10.6 Discussion on configuration pattern study

The business ecosystem configuration pattern study is developed from the following literature: configuration pattern is a synthesised model to

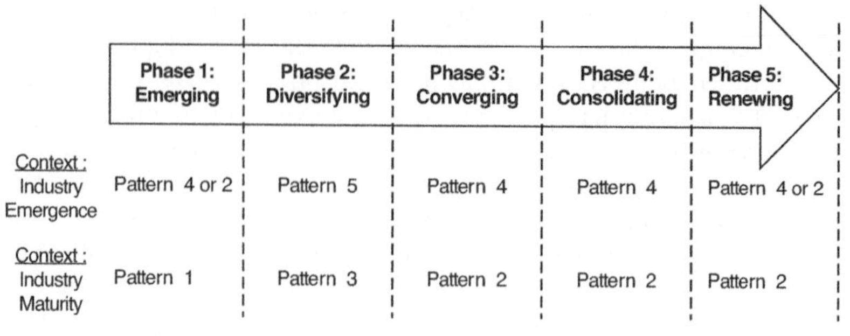

	Phase 1: Emerging	Phase 2: Diversifying	Phase 3: Converging	Phase 4: Consolidating	Phase 5: Renewing
Context : Industry Emergence	Pattern 4 or 2	Pattern 5	Pattern 4	Pattern 4	Pattern 4 or 2
Context : Industry Maturity	Pattern 1	Pattern 3	Pattern 2	Pattern 2	Pattern 2

Figure 10.4   Pattern evolution along business ecosystem life cycle

reflect manufacturing system typical strategy from a specific perspective (Hayes & Wheelwright 1984). For example, the process and product were used to categorise firm level manufacturing systems (Hayes & Wheelwright 1979), and several configuration patterns were identified regarding the different matches between processes and products. This configuration pattern is a synthesised model to demonstrate firm/ manufacturing system's typical strategy and daily operation. Later, in 1998, Shi and Gregory (1998) categorised the international manufacturing network level by using two dimensions: geographic dispersion and horizontal coordination to clarify intra-firm level system collaboration and coordination. Later, other scholars also used similar strategies to categorise a global supply network (Zhang et al. 2007; Srai & Gregory 2008).

Learning from previous theories, the business ecosystem configuration pattern theory demonstrates a synthesised model to easily and precisely understand firms' strategy within their business ecosystems. This theory helps to identify firms' strategy in nurturing the business ecosystem which expands the scope of previous studies. In this study, seven configuration patterns are explored by using the two dimensions of Solution platform openness and Solution diversity to demonstrate different typical ecosystem strategies in order to cope with the two industrial requirements of interoperability and uncertainty. Furthermore, this study also highlights the configuration pattern evolution model along the BELC.

In summary, there are three advantages of the business ecosystem configuration study: first, as categorised by the most representative constructive elements, configuration patterns are able to demonstrate the key features of the business ecosystem itself; second, different patterns naturally indicate different strategies of how firms nurture their business ecosystems; third, the pattern evolution model also give firms reference on how to use any of those patterns along the BELC to benchmark performance.

10.7   Conclusion

This chapter extends the theory on the configuration pattern of business ecosystems with four research findings.

First, two dimensions – solution diversity and platform openness are identified to categorise configuration patterns. These two dimensions come from the requirements analysis of two drivers from the industrial background: uncertainty and interoperability. These requirements also reflect business ecosystem constructive elements: solution platform and adaptive solution. As a result, these two factors are the most critical and representative to categorise the configuration patterns of the business ecosystem.

Second, based on the two dimensions, seven out of nine patterns are identified by analysing the data. They are scarce solution ecosystem, platform enabling ecosystem, platform integrated ecosystem, platform coordinating ecosystem, platform co-evolving ecosystem, open community ecosystem coordinated by few companies and open community ecosystem. All the seven patterns demonstrate two factors: power distance and innovation scope. Platform openness reflects the power distance between partners, or entry barrier, while solution diversity means the quantity of different organisations who contribute towards innovation. Patterns 1 to 5 work with business orientation, while Patterns 6 and 7 usually are run by the open community without business purpose. However, in order to share resource with both sides, companies in the business side and volunteers in the community side will interact with one another to transfer the value and opportunities.

Third, three main cases' general strategies on the pattern evolution are developed. ARM's strategy is very open and flexible. Intel has begun to open its Solution platform in the mobile computing industry. MTK dominates the market by introducing the Turnkey solution with low cost and low entry barrier. However, on some occasions they also accepted ideas from the partners in order to enrich the platform and maintain their competitive advantage.

Fourthly, based on the evolution of the three cases, two types of general configuration pattern evolution are identified. In the context of mature industry, firms follow the path of Pattern 1, 3, 2, 2 and 2 at the related five phases. In emerging industry, firms follow Pattern 4 or 2 in Phase one, Pattern 5, 4, 4 in the following three phases and Pattern 4 or 2 in the final phase.

Finally, a discussion on the study of configuration patterns was conducted by comparing the studies at firm, intra-firm, inter-firm, internationalised inter-firm and business ecosystem level.

11
Business Ecosystem Nurturing Process

11.1 Introduction

The five-phase business ecosystem life cycle (BELC) has been developed as well as the ecosystems' construct and configuration patterns along the life cycle. This chapter especially addresses the question of how firms nurture business ecosystems under the life-cycle stages: how they deal with the business ecosystems' constructive elements and develop different configuration patterns. What are the key constructive elements out of those eight elements during the nurturing process?

11.1.1 Nurturing process: firm, business ecosystem and industry

Specifically, the life-cycle study demonstrates a business context for firms' activities which includes five phases: Emerging, Diversifying, Converging, Consolidating and Renewing. This chapter will explore in detail along those five phases the process of how firms nurture their own business ecosystems from the firm's perspective.

It is very important to further explore the process of firms nurturing business ecosystems for the following two reasons: first, life-cycle studies only focus on the business context level which could not comprehensively explain the details of the growth process and the internal operational mechanisms of the business ecosystem. Secondly, the study of a firm's nurturing process would explore the interaction among partners along the BELC. More specifically, this study shows the process of how a firm uses its business ecosystem constructs to develop not only a single product but also generations of products to sustain the commercialisation process for the emergence and further development of new industries.

In other words, the BELC is the context in which firms engage with their partners for new industry development. In short, firms use the business ecosystem to develop new industries.

11.1.2   The methods of conducting research for nurturing process

Each phase of the BELC has its own characteristics and a different business context. As a result, in terms of the unique features of each phase, individual firm's nurturing processes would vary. Thus, this chapter looks at the firm level by studying those main cases and their activities along the BELC in order to explore the logic among the firm, business ecosystem and specific industry.

From the research methodology perspective, the case data in this chapter is abstracted from Appendix Figures 1, 2, 3. The study focuses on the firm's sequential projects which show firms nurturing their business ecosystems through each sequential phase of BELC as in Figure 11.1: horizontally, the study will be conducted on each main case, project by project, to explain their own unique nurturing process along the five phases of the life cycle; vertically, this study will conduct cross-firm analysis and finally deliver the general nurturing process along BELC.

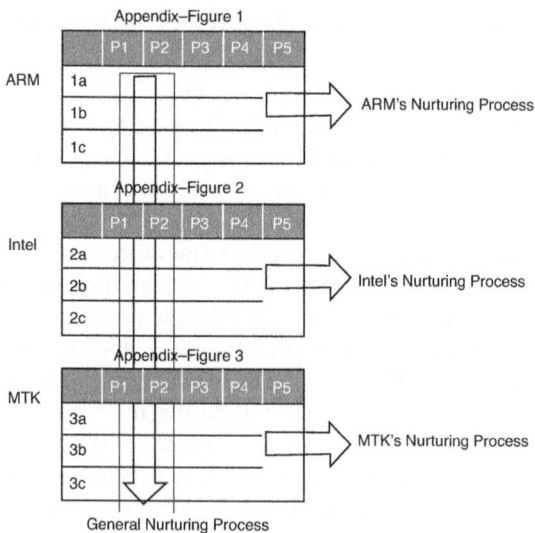

Figure 11.1   Methods of developing nurturing process

11.2 Three cases' typical nurturing process analysis

11.2.1 Data links

This section aims to develop the nurturing process from the perspective of each main case. By learning from Appendix Figures 1, 2 and 3 separately, each main case's activities along the five phases of the life cycle will be identified.

11.2.2 The pattern ARM's nurturing process

Table 11.1 shows the ARM case abstracted from Appendix Figure 1. From the particular study about ARM's business ecosystem growth, some key points can be highlighted as well as ARM's strategies for nurturing ecosystems.

1) To introduce new ideas to create new market: Low power IP and its IP business model was launched by ARM in the mobile market and this drove specialisation in the semiconductor supply chain and also stimulated the growth of the mobile market.
2) To leverage the mobile phone supply chain: The IP business model was regarded as the turning point in driving specialisation of the mobile phone industry and enabling firms' innovations. Furthermore, as a very small company, ARM persuaded Nokia and built up a new supply chain that also triggered the other partners' contributions.
3) To implement leader partners strategy: Learning from the mobile phone project, ARM realised the importance of chip design firms. They started to conduct a leader partners strategy to find top design players in order to jointly promote ARM's new generations of IP. This strategy enabled ARM to attract many top players and their own business ecosystem partners. Finally, ARM was able to maintain its competitive advantage in the relevant markets.
4) To enable the variety of market: In order to encourage partners' contribution, ARM implemented numerous strategies to support both ISV, OSV and foundry players who could work closely with ARM's leader partners. To all of them, ARM built up the ecosystem and allowed them to work on the proposed IP platform. All could be very easily accessed by ecosystem partners.
5) To optimise its business ecosystem: As ARM's ecosystem became stronger, they had to consider the specialisation of the market requirement, so ARM tailored their IPs and made them compatible with the specialised markets as well as re-organising their business ecosystem partners.

From the above five key points and their detailed processes which are listed in the Table 11.1, the key patterns of nurturing process along the BELC have been concluded to be:

Phase 1

Three key features. First, ARM introduced the new ideas (low power for future mobile) to encourage partners to develop a future mobile phone. Then ARM developed the low power solution and the relevant IP license business model to attract partners' participation (in this stage, ARM only regarded its low IP as a product; in the next phase, ARM realised it could act as the product platform to integrate downstream partners). Thirdly, ARM leveraged the supply chain by coordinating more partners beyond the direct supplier and customer.

Phase 2

Four key characteristics are highlighted. First, ARM started to know the low power IP was not only an incomplete product, but also able to act as the product platform to encourage the downstream partners' involvement. Then ARM started to introduce the product platform. Secondly, ARM is to co-design the products by collaborating with leader partners; the third is to co-design the future industry picture (vision) with partners. The last one is to support partners and enable the network and products to diversify through flexible collaborations based on the product platform.

Phase 3

In this phase, ARM began to identify the specialised markets and encourage ecosystem partners to select the end-user solutions. Then ARM categorised and tailored the IP platform to adapt to those specialised market requirements. Besides, ARM also set up alliances for promotion of those products. As a result, the partners were also taking up specialised positions.

Phase 4

In order to scale up market volume and improve industry efficiency, ARM continued to improve its IP platform and consolidate the end-user solutions as well as closely supporting its fixed partners.

Phase 5

This phase is the key step to renew the existing market with niche ideas. So in this phase, an upgraded idea or solution helps the old industry to step into the new ones. Then this phase connects with Phase 1 and subsequently re-starts the BELC.

Table 11.1 ARM's nurturing process

			Business ecosystem life cycle		
Phase **Projects**	**Phase 1** **Emerging**	**Phase 2** **Diversifying**	**Phase 3** **Converging**	**Phase 4** **Consolidating**	**Phase 5** **Renewing**
1a- Mobile phone	1. ARM shared low power vision for future mobile 2. ARM developed IP with low power 3. Leveraged partners to create new supply chain.	1. Co-designed the product and future mobile vision 2. ARM supported software partners. 3. ARM shared IP with design tools and manufacturing partners	1. ARM finalised the low power vision for mobile 2. ARM's solution followed by many OEMs 3. ARM built up the strong alliance	1. ARM9 was improved from ARM7 2. ARM continued to cooperate with TI and Nokia 3. ARM also supported software firms	1. ARM7 was re-designed for embedded market
1b- Leader partner strategy (LPS)		1. ARM recycled ARM7 and developed it into Cortex M3 2. Co-marketed as LPS 3. ARM built up connected community	1. Identified the priority of embedded market 2. ARM and lead partners formed small alliance for new IP promotion 3. Succeeded ARM7 ecosystem	1. ARM worked with more LPS 2. More than 100 customers chose SMT32 (ARM-based chips)	
1c- IP categorisation			1. Identified three streams market 2. ARM customised IPs into Cortex A, R, M series 3. Connected community support	1. Continued LPS 2. Those IPs were improved to approach best solution 3. The previous partners were reorganised	
Conclusion ARM's nurturing process	1. Share low power vision. 2. Introduce New solution for Emerging industry 3. Leverage partners for building supply chain	1. Collaborate with lead partners 2. Co-design the future vision with partners 3. Make solution diversity through highly flexible collaboration 4. Introduce low power IP platform	1. Identify market specialisation 2. Close alliance with key partners 3. Select solution well accepted 4. Categorise platform	1. Continue to improve solution 2. Supports fixed partners 3. Consolidate the platform	1. Upgrade solution

Source: Concluding from Appendix Figure 1.

In summary, ARM was acting very openly by involving many leader partners and supplementary partners and their contributions to nurture its business ecosystem. ARM's leader partners' strategy was one of the key tools to introduce its product platform and develop ARM's ecosystem with great support from those top players as well as their partners.

Figure 11.2 visualises the nurturing process from ARM's perspective.

11.2.3   Intel's nurturing process pattern

Table 11.2 is abstracted from Appendix Figure 2, the detailed Intel case. From the particular study concerning the growth of Intel's business ecosystem, some key points can be highlighted, while Intel's strategies on nurturing ecosystems are also worth mentioning.

1) To introduce industry standard

Intel won the dominant design role as they introduced the PCI interface as the industry standard interface. This industrial standard enabled partners with supplementary products to work easily with Intel's processors and deliver the end-user products. Furthermore, through this interface, partners' products would be compatible with generations of Intel's processors.

2) The failure of Xscale project

The key reason that Intel failed in the Xscale project was that Intel did not nurture and attract partners but just closely collaborated with Motorola. As a result, there were few partners supporting Intel's solution for mobile phones and their simplified end-user products. To make the matter worse, Intel also faced intense competition from top players in the mobile phone market. In summary, this case indicates that it is

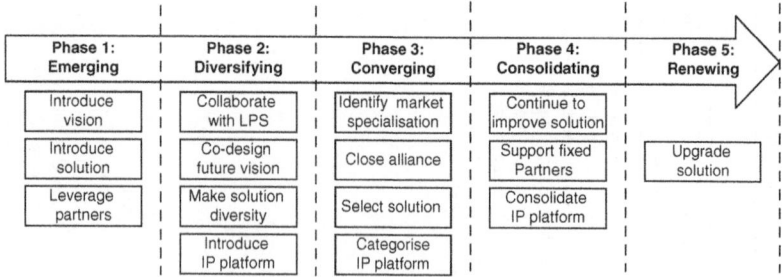

Figure 11.2   ARM's pattern to nurture business ecosystem

important to nurture partners' network and their contribution in order to accomplish solution diversity.

3) Learning from partners' idea – EeePC

A very wonderful idea of 'good enough' was adopted by Asus who used Celeron chips to introduce the low cost notebook EeePC. It was a niche idea to renew the traditional notebook market. Finally, Intel also persuaded the other partners to contribute to this niche market.

4) Netbook as a start point of mobile computing strategy

The Netbook idea was learned from EeePC, and Intel regarded this opportunity as a chance to re-enter the mobile computing industry. Therefore the concept of Netbook was finalised differently from note-book. They succeeded in persuading other partners to enable market variety. With the strong background of notebook experience, this new industry matured very quickly and the product specification was also finalised as an industrial standard.

5) Enable variety through mobile ecosystem and MID project

These two projects were aimed at bringing out new ideas or solutions to enable contribution from partners. The mobile ecosystem was operated as an open source community and MID was co-designed by partners. The software from the mobile ecosystem is also supplementary to the MID project.

From the above four key points and their detailed processes as listed in Table 11.2, the key pattern of the nurturing process along the BELC has been concluded to be:

1) Phase 1

First, Intel shared its vision of the development of a new industry (specialised trend of the PC industry and low power for the mobile computing industry). Then Intel introduced their new solutions based on their own market (PC industry) and their partners (EeePC). Finally Intel would persuade their previous partners to enter this new industry or to initiate a new supply chain.

2) Phase 2

In this phase, Intel's strategies became open rather than only supporting many OEMs: they initiated the concept of a future industrial product

Table 11.2 Intel's nurturing process

Phases project	Business ecosystem life cycle				
	Phase 1 Emerging	Phase 2 Diversifying	Phase 3 Converging	Phase 4 Consolidating	Phase 5 Renewing
2a PC	1. Supported specialised PC industry 2. Provided CPU processors 3. Intel was key supplier of IBM's network	1. Intel was supporting different industry architecture 2. Those OEMs used Intel's processor to differentiate their products	1. Intel discovered it was very difficult to support so many partners 2. Initiated PCI platform 3. Used PCI coordinate partners	1. Intel continued to introduce standards: USB, Wimax 2. Intel developed more advanced technology 3. Intel set up stable partners' network with Microsoft	1. Intel decided to enter mobile by ARM IP 2. Intel encouraged PC partners enter into mobile world
2b Xscale	1. The low power solution from ARM 2. Intel tried to persuade their PC partners to adopt its new processors for mobile phone		1. Intel just launched close alliance with Motorola in order to reach the dominant design	1. Intel also outsourced manufacturing to TSMC in order to reduce the lead time	1. High cost and low support 2. Intel sold the Xscale to Marvel
2c Atom	1. Intel pushed the low version of notebook 2. They developed Atom as a new low power platform 3. Succeeded from PC industry partners	1. Intel initiated the concept of Netbook and MID 2. Intel also provided much financial support to partners 3. Intel also built up Moblin ecosystem	1. Intel introduced Netbook industry standards 2. Intel selected key OEM to develop MID 3. Intel kept improving its Atom solutions and software		
Conclusion	1. Share vision of specialised PC industry 2. Introduce New solution 3. Encourage partners & support simple supply chain	1. Co-design product concept 2. Initiate Software ecosystem 3. Support different OEMs 4. Introduce software and hardware platform	1. Introduce industry standard 2. Select Partner network 3. Improve solution 4. Improve the platform	1. Continue to introduce industry standard 2. Keep improve advance technology. 3. Set up stable partner network 4. Consolidate the platform	1. Introduce new idea 2. Re-organise partners or 1. Project failure and sold out

Source: Abstracted from Appendix Figure 2.

by using its hardware platform, for example Atom; they also built up open source software as part of the product platform as well as providing financial support.

3) Phase 3

In this phase, Intel very successfully won competitive advantage not only in the PC industry but also in the Netbook market as they introduced industry standards accepted by their partners. Based on those industry standards, Intel also selected partners to develop new products. Intel continued to improve their product platform (processors) to further strengthen its end-user solution performance.

4) Phase 4

In order to maintain competitive advantage and improve industry efficiency, Intel continued to introduce new industry standards such as USB and Wimax. Besides this, Intel also kept improving the advanced technology used to manufacture their processors. Finally, Intel set up a stable partner network to achieve the dominant design. Thus combination also seemed as the platform consolidation for the industries end-user solution.

5) Phase 5

In this phase, Intel inserted many niche ideas as the current industry became saturated. They also tried to persuade their previous partners to create the new industry together. However, Intel also experienced failure as they could not maintain the Xscale project.

In summary, Intel introduced industry standards to coordinate and control partners' activities in order to improve industry efficiency. However, they experienced failure when they just copied those strategies and failed to nurture their partner network in the Xscale project. They learned from this failure that it is very hard to control industry development simply by introducing industry standards when the industry becomes very complex. As a result, Intel started to co-evolve with their partners and initiated a software community and hardware platform to attract more partners to work together.

Figure 11.3 visualises the nurturing process from Intel's perspective.

11.2.4   MTK's nurturing process pattern

Table 11.3 is abstracted from Appendix Figure 3, the detailed MTK case. From the study of the growth of MTK's business ecosystem, some key

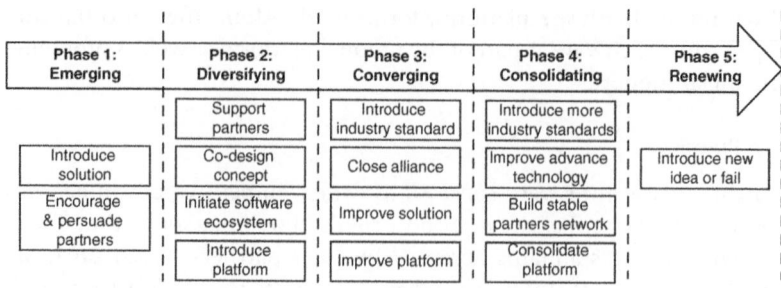

Figure 11.3   Intel's pattern to nurture business ecosystem

points as well as Intel's strategies of nurturing the ecosystem can be highlighted.

1) The Turnkey model

MTK's Turnkey model was widely implemented in VCD, DVD, mobile and smartphone projects. Furthermore, the Turnkey model helped cut down the industry barriers and decreased the production cost.

2) Enable downstream innovation

Besides the Turnkey model, MTK also exploited the strong manufacturing capability of their partners' network in mainland China. MTK very successfully triggered their innovation activities and successfully delivered diversified products with low cost.

3) Very quick to produce the dominant design

After the industry reached a very significant volume, MTK aimed to produce the dominant design with user-friendly features and low cost. This strategy beat down many big players in the industry.

4) Co-evolve with local partners

As more requirements emerged from marketing, MTK also co-evolved with local requirements and added more functions to the Turnkey model as well as supporting local partners.

5) Expand partners scope

There are more complicated networks in the smartphone industry. MTK not only collaborated with previous partners but also built up relationships with carriers and design firms as well as contributors to the Android open source community.

From the key points above and their detailed processes listed in the Table 11.3, the key pattern of nurturing process along the BELC is concluded to be:

1) Phase 1

At first, MTK initiated the concept of a single-chip solution to cut down cost and industry entry barriers. Then, MTK not only provided the single-chip solution for mobiles, but also brought in a new pattern of partners' interactions to enable innovation from downstream firms in the supply chain. Following the Turnkey model, MTK could easily get the supply chain to speed up the commercialisation process.

2) Phase 2

In this phase, MTK introduced the Turnkey model and invited partners' contribution to product design. This platform could enable solution variety by co-evolving with downstream local partners and adopting their feedback. MTK successfully triggered innovation within the Shenzhen area. MTK also provided training sessions to make it easier for engineers to work with their Turnkey model.

3) Phase 3

MTK's method in this phase was to highlight the importance of approaching the best solution for the market requirement by further improving its Turnkey platform. Hence, in order to improve industrial efficiency, they also selected partners to work with. By learning from the second phase of solution diversity, MTK in this phase decided to deploy their products into several streams.

4) Phase 4

In this phase, in order to maintain competitive advantage, MTK started to consolidate its product platform by building up very close alliances with its local partners. Besides this, the improved platform also secured the industrial dominant design.

5) Phase 5

In this phase, MTK also introduced new ideas to renew the existing industry such as renewing VCD to DVD and 2G to smartphone. In order

Table 11.3 MTK's nurturing process

	Business ecosystem life cycle				
Phases project	Phase 1 Emerging	Phase 2 Diversifying	Phase 3 Converging	Phase 4 Consolidating	Phase 5 Renewing
3a VCD/DVD	1. MTK initiated single-chip solution with low cost 2. This solution cut down high technology barrier	1. MTK sold Turnkey model to various partners in Hsinchu park 2. MTK provided the software design kit	1. MTK followed OEM to mainland China 2. MTK also enabled local OEMs innovation in Shenzhen	1. MTK still kept its solution with low cost and low entry barrier 2. MTK continued to improve its solution into the speed of 24X	1. MTK upgraded VCD into DVD 2. MTK also succeeded the VCD partners' network
3b Mobile 2G	1. Concept of single-chip solution 2. Provided Turnkey model of mobile phone solution 3. Collaborated with a small local Taiwan OEM	1. They finally enabled Shanzhai manufacturing network innovation 2. Made the solution diversity	1. MTK supported independent design house partners 2. MTK also did much training for local design engineers	1. MTK continued to improve Turnkey model as they added more novel function into the solution 2. Finally 20% China shipment	1. MTK introduced low cost of smartphone solution 2. Success from 2G mobile network
3c Smartphone		1. MTK introduced smartphone succeeding from 2G firm network 2. MTK collaborated with hardware, software, design partners 3. MTK also provided design support and training session	1. MTK finalised their design with two streams 2. MTK selected partners		
Conclusion	1. Share vision of single-chip solution 2. Introduce Turnkey solution 3. Initiate the new supply chain	1. Co-design Turnkey with more partners 2. Co-evolve with different partners 3. Design and training support 4. Introduce Turnkey as the platform	1. Select specific partners of Shanzhai network 2. Deployed their product streams 3. Improve the Turnkey platform	1. Close alliance with partners 2. Improved end-user solution as the dominant design 3. Consolidate the Turnkey platform	1. Introduce new idea 2. Succeed previous network

Source: Abstracted from Appendix Figure 3.

to cope with new market requirements, MTK also reorganised their partners' network.

In summary, MTK likes to penetrate a market when it is very mature and of high production volume. Usually they use a Turnkey solution to enable thousands of small firms to deliver the products with low cost and short lead time. However, when they entered the emerging and complicated mobile computing industry, they began to co-evolve with more types of firms than before, whose feedback they embedded. Besides this, MTK also contributed to the open source community, provided design support and training sessions for the design engineers.

Figure 11.4 visualises the nurturing process from MTK's perspective.

11.3　General nurturing process along the business ecosystem life cycle

11.3.1 Data links

In the last section, three patterns of nurturing processes from the perspectives of each main case have been developed. In this part, cross-firm analysis will be conducted as in the vertical axis of Figure 11.1 in order to deliver a general nurturing process along the BELC. The projects are presented in detail in Appendix Figure 1 (ARM), Appendix Figure 2 (Intel) and Appendix Figure 3 (MTK), which were integrated as the general nurturing process in Figure 11.5. Regarding each phase, the definition of sequential nurturing steps has been identified in Table 11.4. After that, conclusions about the detailed processes along the BELC are drawn in the following section.

11.3.2　Nurturing process in emerging phase

First, on one hand, the firm has to persuade potential partners by sharing their future vision of the industry (like low power for mobile

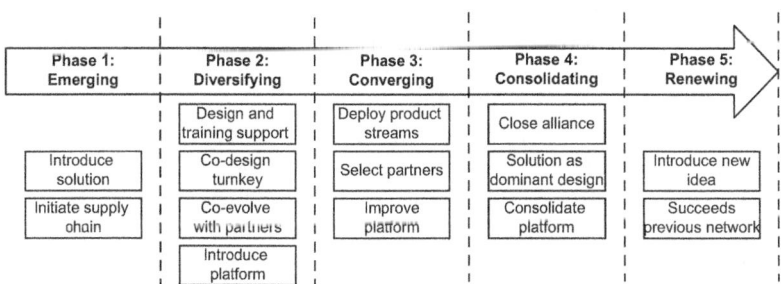

Figure 11.4　MTK's pattern to nurture business ecosystem

phone market); on the other hand, the firm also introduces some new solutions for the emerging industry like ARM's IP for mobile phone and Intel's processors for the PC industry. Secondly, besides the new solutions, firms also leverage original or new partners to work together. For example, ARM initiated the consortium with NOKIA and TI while Intel persuaded their PC partners to contribute to the mobile market. Their aims are to build a new supply chain by leveraging partners' positions and contributions. All in all, this phase demonstrates methods to accelerate the commercialisation of emerging ideas by coordinating different partners and resources and giving birth to new products.

Three steps seem to be important: 'share vision', 'introduce solution' and 'leverage partners':

- 'Share vision': To share its future thoughts of the emerging industry with partners
- 'Introduce solution': To introduce novel solutions for an emerging market either from self-development or adoption from partners
- 'Leverage partners': To coordinate partners along the value chain and to quickly persuade partners in order to build up a supply chain and commercialise products

Table 11.4 lists those definitions of each phase.

11.3.3   Nurturing process in diversifying phase

After the Emerging phase, the most important element in the second phase is to involve more and more partners' contributions to the whole process of new product development along the supply chain. First, the firm co-designs the future vision of the industry with their partners. Second, the firm started to introduce the product platform for end-user solutions. Third, more partners are encouraged to contribute further to the solutions for this industry. For example, ARM introduced the new product platform and offered the opportunities to the leading strategic partner to co-design the future IP, Intel built up the Moblin ecosystem, while MTK joined the Android open source community to encourage partners to co-design the future product. Fourthly, partners' interactions will enable the solution to be diversified in order to meet the industrial requirements and to build up an interface for partners' interaction. For example, OSV and ISV were adding value to the ARM and Intel platforms and then differentiating the end-user products. So with the help of partners' contributions, this phase will achieve solution variety. As a result, it boosted the prosperity of the ecosystem and created more business opportunities to stimulate the companies' passion.

All in all, this phase consists of 'Co-design vision', 'Introduce platform', 'Encourage partners' and 'Enable solution diversity'.

- 'Co-design vision' involves partners in co-designing the Solution platform for the future of the industry.
- 'Introduce platform' is to encourage the partners to develop the end-user solution based on the initiated platform.
- 'Encourage partners' co-designs the image of the future of the industry with partners.
- 'Enable solution diversity' co-designs the end-user product with partners and enables end-user product solution variety based on a specific platform.

11.3.4   Nurturing process in converging phase

This phase focuses on the efficiency of industrial productivity. After various innovative activities in the second phase, the industry vision is finalised in order to select some typical well-accepted solutions to improve industrial productivity and efficiency. For example, in the mobile market, MTK used a single-chip product platform to cut down the cost of mobile phones and stimulate the productivity of the local manufacturing partners' network. Intel also used many industrial standards to finalise industry vision and further improve the product platform with key partners and then to encourage partners to work to the standards. Those who cope with the selected solutions are selected from the business ecosystem as the specific network partners. MTK further improved their own product platform and selected partners who wanted to develop low cost mobile phones and required design support as well. Intel filtered the partners by its standards of PCI interface. As a result, by going through the emerging solutions and various solutions in the first two phases, the third phase aims to reach a typical solution.

All in all, this phase is composed of three steps: 'Finalise vision', 'Select solutions', 'Select partners' and 'Co-design platform':

- 'Finalise vision' improves industrial efficiency by best solution and partner re-organisation.
- 'Select solutions' selects the well accepted end-user product design from the industrial requirements.
- 'Select partners' re-organises the partner network to one which is suitable for the best solution product.
- 'Co-design platform' continues to work with key partners to improve the solution platform.

	Phase 1: Emerging			Phase 2: Diversifying		
1a: ARM7/ ARM9	Share low power vision for future mobile phone market	Low power IP solution; a new license model	Leverage partners to create new supply chain	Co-design the future mobile solution	Co-design with TI and Nokia's feedback	Support other partners to add value
1b: LPS				Co-design the embedded market direction, recycled from ARM7	Leader partner strategy with top players; Co-design and co-marketing	Supporting connected community
1c: IP clarify						
2a: PC	Specialised PC industry	PC processor as the main business	Act as a key processors' supplier to IBM	Collaborate with partner for future product	Intel support different industry architecture from various OEMs	
2b: Xscale		Launch Xscale project for mobile	Try to persuade previous partners			
2c: Atom	Identify new requirement of mobile computing industry	New solution as Atom	Succeeds PC industry	Initiate the idea of Netbook & MID	Atom Platform evolution; Support different partners around Atom platform	Moblin ecosystem for software development for Atom platform
3a: VCD/ DVD	Turnkey solution to cut down industry barrier	Provide Turnkey solution to mainland China			Provide software design kit	Sold Turnkey to various OEM in Hshinchu park
3b: Mobile 2G	Low cost, easy design with	Provide Turnkey model to Shenzhen areas	Collaborate with a small Taiwan OEM		Enable the end-user product variety	Enable innovation in Shanzhai manufacturing network
3c: Smart-phone				Low cost; easy design based on MTK platform with Android community;	Enable non-direct partners' innovation around MTK platform; try to connect Carrier; penetrate into 3G standards	Co-evolve with local partners; design support and training session; join Android community
Detail process	Share vision	Introduce solution	Leverage Partners	Co-design vision	Co-design solution; introduce platform	Encourage partners

Continued

	Phase 3: Converging		Phase 4: Consolidating		Phase 5: Renewing	
Low power solution	Win the competitive advantage followed by OEMs	Strong alliance with TI and Nokia	Continue to improve its IPs	Keep supporting partners with more activities	ARM7 is re-designed for embedded market	Find new partners for the IP
Low power and high performance	Small alliance promote Cortex M3's performance	Strong software ecosystem	More than 100 customers chose ARM based SMT32	ARM works with more Leader partners		
Identify three streams of market	ARM customises IPs into Cortex A, R, M series	Connected community support	Continue LPS those IPs are improved to approach best solution	The previous partners are reorganised		
Difficult to collaborate separately, promote platform based collaboration	Leader platform strategy as PCI	Partners around PCI standard	Continue to introduce standards; more advanced solution	Stable partners' network, maintain the barrier like Microsoft	License Strong ARM from DEC; enter mobile	Encourage partners to enter new market
		Close alliance with Motorola		Outsource chip manufacturing to TSMC	Intel sell Xscale to Marvell	
Atom and its relevant software open source improved	Introduce Netbook industry standards	Select partners with Netbook standard				
Low cost and easy design with	Single chip solution	Work with VCD player and PC players in mainland China	Keep low cost and low entry barrier and reach speed of 24X	Enable the innovation in China	Using DVD to renew VCD	Quickly succeed from previous network
Low cost and easy design with	Support chip design house	Provided training sessions	Approach dominant design with local partners	20% China shipments	MTK introduces low cost model of smartphone	Succeed previous network
	Deploy their smartphone with two 3G standards	MTK selects partners to promote 3G solutions				
Finalise vision	Select solution; co-design platform	Select partners	Improve & Finalise solution; consolidate platform	Integrate partners	Initiate & Capture Niche idea	Reorganise partners

Figure 11.5   Detail nurturing process of each phase

Source: Appendix Figures 1, 2, 3.

11.3.5   Nurturing process in consolidating phase

In the Consolidating phase, some typical end-user solutions selected in the converging phase are finalised by partners. The solutions platforms are polished and consolidated in order to be accepted by most of the partners. For example, ARM continued to improve their IPs and provided more support activities. Intel continued to improve their PC processors and introduced more industrial standards in order to maintain their competitive advantage. MTK won their market share by continually improving the Turnkey model. The Consolidating phase will last for a long time and the partners' network will not change much. Central firms aim to finally integrate partners for the end-user products. In the PC market, Intel and Microsoft have integrated partners along the PC supply chain since the 1990s. MTK triggered the integrated Shanzhai manufacturing network. As a result, this phase finalises the product design and forms the close partner alliance so as to achieve mass production for a long period.

In conclusion, this phase consists of three steps: 'Finalise solution', 'Integrate partners' and ' Consolidate platform':

- 'Finalise solution' continues to improve the end-user solution in order to dominate the market.
- 'Integrate partners' integrates the partners' network around the dominant design in order to improve industrial efficiency.
- 'Consolidate platform' continues to improve the solution platform in order to maintain the competitive advantage and lock in the partners.

11.3.6   Nurturing process in renewing phase

The last phase aims to find new ideas in order to prevent the obsolescence of the business ecosystem. On the one hand, the key market situation with niche requirements has been identified. Then the new idea is initiated either by central firms or partners to cope with a niche market requirement as the original industry becomes saturated. For example, Intel hoped to use ARM's IP to renew their PC business ecosystem by producing mobile devices. MTK used DVD chips to renew the mature VCD market. In addition, new ideas also help the companies to change the network organisation – the way that the industrial players interact. MTK's Turnkey model trigged the downstream innovation in the mobile market, which was hugely different from the original working of the close alliance. Intel's Mobile project also changed itself to be more open;

for example, Intel outsourced the manufacturing of ARM-based chips to TSMC. After that, companies renew the ecosystem internally by re-entering the first phase again – but there are now plenty of partners and resources for companies to quickly build up a new supply chain. This phase demonstrates the way that firms bring in niche solutions to restart the life cycle. On the other hand, if the niche ideas do not emerge and are not captured, the original industry will decline.

All in all, this phase is composed of two steps: 'Initiate and capture niche idea' and 'Reorganise partners':

- 'Initiate and capture niche idea' identifies the new requirement/ niche market which is different from the existing industry.
- 'Reorganise partners' provides the way to change partners' network for niche market.

11.3.7 Synthesis of nurturing process along business ecosystem life cycle

Table 11.4 summarises the detailed nurturing processes in each sequential phase of the BELC. By synthesising the research findings in Chapter 5 and Chapter 6, Figure 11.6 visualises a five-phase BELC and nurturing processes in each phase.

In phase one, Emerging, firms start to share vision and introduce new solutions to persuade partners to work together. As a result, a supply chain is initiated for this new solution. In phase two, Diversifying, firms co-design an industry vision with partners, introduce the product platform and encourage more partners to enable solution diversity. As a result, diversified solutions are introduced and many partners are interacting with one another for those solutions. In phase three, Converging, firms begin to finalise the industry vision and further improve the product platform with key partners and enable them to select some typical end-user solutions that meet the market requirements. As a result, partners will be selected and the market will become specialised. In phase four, Consolidating, the central firm continues to consolidate the product platform to lock in the partners in order to maintain competitive advantage. So the central firm will integrate its partners in order to improve industrial efficiency and achieve mass production. In phase five, Emerging, firms will introduce some niche ideas in order to persuade partners to enter another relevant emerging industry. The partners' network will be reorganised in terms of the niche ideas. Then the business ecosystem enters phase one again to leverage the partners

Table 11.4   Nurturing steps identification

	Sub-dimensions	Description
Phase 1	1. Share vision	To share its future thinking of the emerging industry with partners
	2. Introduce solution	To introduce the novel solutions for emerging market either from self-development or adopting from partners
	3. Leverage partners	To coordinate partners along value chain and fast persuade partners to build up supply chain and commercialise products
Phase 2	1. Encourage partners	To involve partners to co-design the solution platform for industry future
	2. Enable solution diversity	To co-design the product with partners and enable the products solution variety based on specific platform
	3. Co-design vision	To co-design the new image of industry future with partners
	4. Introduce platform	To encourage the partners to develop the end-user solution based on the initiated platform
Phase 3	1. Select solution	To select well-accepted products' design from industry requirement
	2. Select partners	To re-organise the partner network to suitable for best solution product
	3. Finalise vision	To improve industry efficiency by best solution and partners re-organisation
	4. Co-design platform	To continue work with key partners to improve the solution platform
Phase 4	1. Improve and finalise solution	To continue improving the solution to dominate the market
	2. Integrate partners	To integrate partners' network regarding the dominant design in order to improve industry efficiency
	3. Consolidate platform	To continue to improve the solution platform in order to maintain the competitive advantage and lock in the partners
Phase 5	1. Initiate and capture niche idea	To identify the new requirement/niche market different from existing industry
	2. Re-organise partners	To provide way to change partners' network for niche market

to quickly commercialise the new ideas. However, if none of the niche ideas are initiated and captured, the original industry would decline instead. The BELC consists of these five phases which make for a sustainable growth and transformation of the business ecosystem.

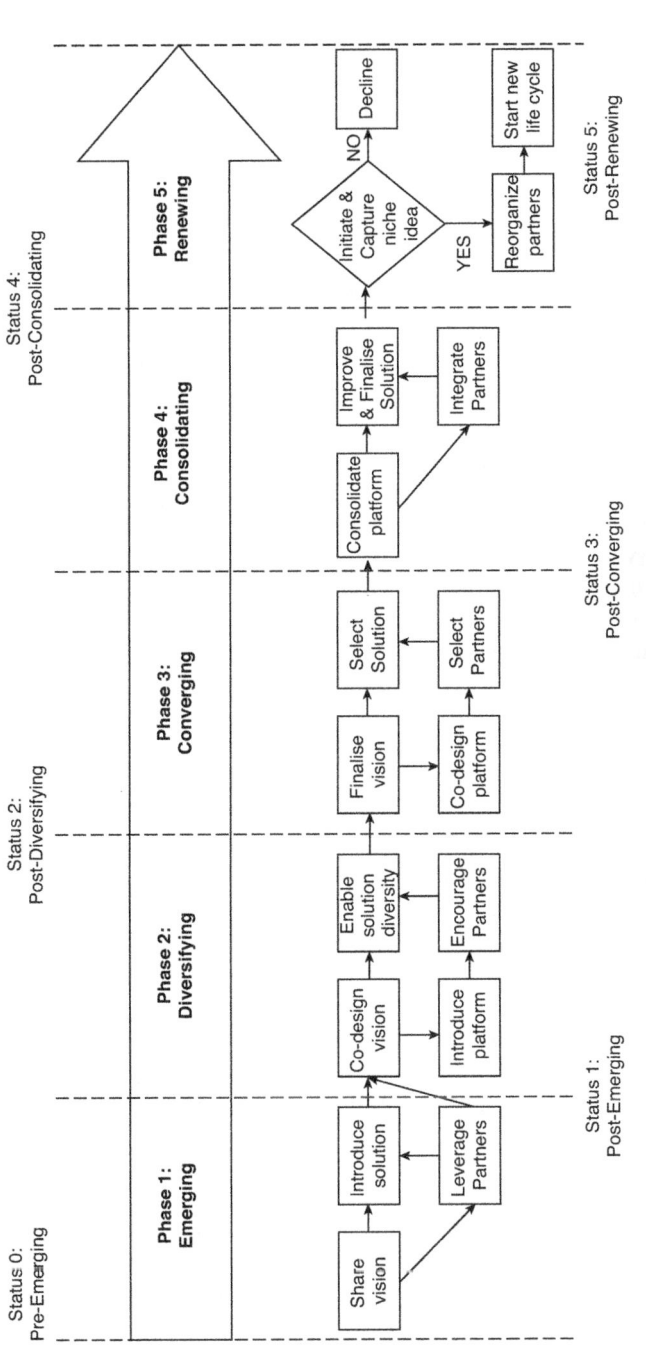

Figure 11.6 Business ecosystem life cycle and its nurturing process

11.4   Nurturing process for industrial practice

The previous section discusses the general nurturing process along the BELC. However, in a real situation, a firm would have a different starting point to nurture its business ecosystem. This part summarises each case's detailed process and delivers three different patterns of nurturing process for practical use.

11.4.1   Projects formulating three patterns

In Appendix Figure 1, ARM's three projects described three different situations and patterns from which to start nurturing a business ecosystem. In the mobile phone project (1a), ARM nurtured its mobile phone business ecosystem **from the beginning**. In the leading partners' strategy (1b), as ARM had already started its business ecosystem in the mobile phone market, ARM could develop the new solution with the support of the existing business ecosystem even though it was very small. This situation is regarded as '**from small to large**'. The key step of this pattern is to enable the solution diversity and partners' contributions. At the start of IP categorisation (1c), ARM already had a strong business ecosystem. However, ARM's solutions were not very customised as performance had been their priority. So ARM had to customise their IPs in order to cope with the specialised market. As a result, this project assisted ARM to improve their business ecosystem **from complicated to optimised**. This project started at the third, or Converging, phase.

In summary, those three patterns of nurturing process demonstrate different strategies with different starting and end points:

- From the beginning: A firm starts to nurture its business ecosystem from the beginning. It starts from the phase of Emerging and proceeds to Diversifying.
- From small to large: after pilot projects, a firm has already built up a partners' network and aims to enlarge the scope of its business ecosystem. It starts from the phase of Diversifying and proceeds to Converging.
- From complicated to optimised: after experiencing a big growth, a business ecosystem becomes very complicated with different kinds of partners and business opportunities. Central firms aim to optimise their partners' network with regard to the specialised market requirements. It starts from the phase of Converging and proceeds to Consolidating.

In summary, ARM's three cases indicated three different points of nurturing business ecosystems. However, those three cases were not only based in one industry. So it would be better to conduct a cross-case analysis within one industry business ecosystem in order to generalise such practical guidance for the nurturing process in a specific industry. In a real situation, different firms, either early movers or late-comers, penetrated into an industry ecosystem with different degrees of industry maturity. It will be of significant practical guidance for those firms to penetrate into the same industry's business ecosystem with different industry development phases. Thus, the firms have to make good use of their existing resources to deal with such various industry maturity.

Within the case database, there are three projects including ARM(1a), Intel(2b) and MTK(2b) based in 2G mobile phone industry, where those firms started to nurture their own business ecosystem from different points of industry development phases. In regard to ARM(1a) project in 2G mobile phone industry, ARM was the key player to bring the IP license model at the very early stage of the mobile phone industry. Now, ARM is the dominating and leading role in this industry development, with 98 per cent of worldwide market share. Thus it is fair to say that ARM's firm-based business ecosystem is equivalent to the whole 2G mobile phone industry business ecosystem. Thus, ARM started the industry business ecosystem from the beginning. Intel penetrated into the mobile phone business ecosystem using Xscale project at the Diversifying phase, which enabled the ecosystem from small to large, while MTK entered at later phase of Converging of the mobile phone industry ecosystem, which enabled the ecosystem from complicated to optimised. Intel and MTK adopted different strategies to nurture their firms' business ecosystem.

In terms of from 'small to large', there are two cases: TI (as ARM's leader partner) was very successful while Intel failed. As the leader partner of ARM, TI had the similar situation of Intel's Xscale case. However, TI's mobile chip was very successful while Intel's failed. One of the key reasons was that TI adopted a co-evolution strategy with partners such as ARM, Nokia and other software partners. TI did not control the end-user solutions design but communicated well with upstream and downstream partners. In contrast, Intel only worked with Motorola and failed to enable the solution diversity. As a result, they finally lacked of partners' support and was failed in this project. As a result, it was important to co-evolve with partners and enable the solution diversity in the pattern of 'from small to large'.

In terms of 'from complicated to optimised', MTK has generated a very efficient way to encourage partners to work around their product platform. Since the industry became mature, the main focus of the business competition shifted from product innovativeness towards production efficiency. Thus MTK provided the Turnkey solution platform which enabled downstream partners to develop the new end-user product very easily at a very low cost. By providing this integrated platform, MTK quickly set up their business ecosystem and succeeded.

As a result, the projects of the three main cases can be placed as three different patterns in Table 11.5 with different starting points.

11.4.2   Industrial guidance

Three patterns of practical guidance can be drawn from Table 11.5 for industrial players who want to build up their own business ecosystems. Those three nurturing patterns are shown in a matrix with two dimensions: time and growing scope of business ecosystem as Figure 11.7.

Pattern 1: If a firm would like to penetrate into a new industry without previous experience, this firm could share their industry's future vision with partners and leverage partners to build up supply chain. They have to deliver the first end-user solution. When the firm has finished initiating its business ecosystem, they will enter the next circle as Pattern 2 as shown in Figure 11.7.

Pattern 2: The key firm has learned from pattern 1 and then provided opportunities to partners to co-design the vision. So as to achieve those visions, the firm should then provide the product platform to encourage partners to work together. It is not like pattern 1 which only provides end-user product, instead pattern 2 provides the platform and end-user product as well. Because the firm wants to enlarge its business ecosystem in some specific industry, this firm should start to co-evolve with its partners and enable solution diversity. Thus, the product platform is the space for those partners' co-evolution and enable solution diversity. This cycle will continue to deliver better accepted solutions. When the firm makes its business ecosystem very diversified, they will enter the next circle as pattern 3.

Pattern 3: If a firm wants to optimise its business ecosystem due to industry maturity, it is better to finalise the vision and integrate the partners' contribution. Hence, the firm would provide the Turnkey platform to simplify the new product development process and enable the downstream partners' network innovation. So the central firm will finally integrate the partner network and deliver the products with low cost and in a short lead time. The cycle will not stop until they become finalised products which are dominant in the market.

Table 11.5 Different patterns of nurturing process

		Industrial level's Business ecosystem life cycle			
Phase **Projects**	**Phase 1 Emerging**	**Phase 2 Diversifying**	**Phase 3 Converging**	**Phase 4 Consolidating**	**Phase 5 Renewing**
Pattern 1: from the beginning: (1a- mobile phone 2G ARM = Succeeded; TI as the first leader partners for mobile IPs)	1. ARM shared low power vision for future mobile 2. ARM developed IP with low power 3. leveraged partners to create new supply chain	1. co-designed the product and future mobile vision 2. ARM supported software partners 3. ARM shared IP with design tools and manufacturing partners	1. ARM finalised the low-power vision for mobile 2. ARM's solution followed by many OEMs 3. ARM built up the strong alliance	1. ARM9 was improved from ARM7 2. ARM continued to cooperate with TI and Nokia 3. ARM also supported software firms	1. ARM7 was re-designed for embedded market 2. ARM continue work with TI as the key leader partner for mobile IPs
Pattern 2: from small to large : (2b-Xscale for mobile 2G: Intel = Failed)		1. Intel did not intend to make the solution diversity	1. The low-power solution from ARM 2. Intel tried to persuade their PC partners to adopt its new processors for mobile phone 3. Intel just launched close alliance with Motorola in order to reach the dominant design	1. Intel also outsourced manufacturing to TSMC in order to reduce the lead time	1. High cost and low support 2. Intel sold the Xscale to Marvel
Pattern 3: from complicated to optimised (3b-mobile 2G: MTK = Succeeded)			1. MTK provided Turnkey model to cut entry barrier 2. MTK supported local Shanzhai network such as independent design house partners 3. MTK also did much training for local design engineers. 4. They finally enabled Shanzhai manufacturing network's innovation	1. MTK continued to improve Turnkey model as they added more novel functions into the solution 2. Finally 20% China shipment	1. MTK introduced low cost of smartphone solution = Succeeded from 2G mobile network
Nurturing steps (Pattern 1)	1. Share vision 2. Introduce end-user solution 3. Leverage partners				
Nurturing steps (Pattern 2)		1. Co-design the vision 2. Co-evolve with partners 3. Enable the product diversity 4. Introduce platform			
Nurturing steps (Pattern 3)			1. Finalise the vision 2. Improve the platform as Turnkey model 3. Enable the downstream network innovation 4. Simplify the solution (easy to deliver)		

Source: Appendix Figures 1, 2, 3.

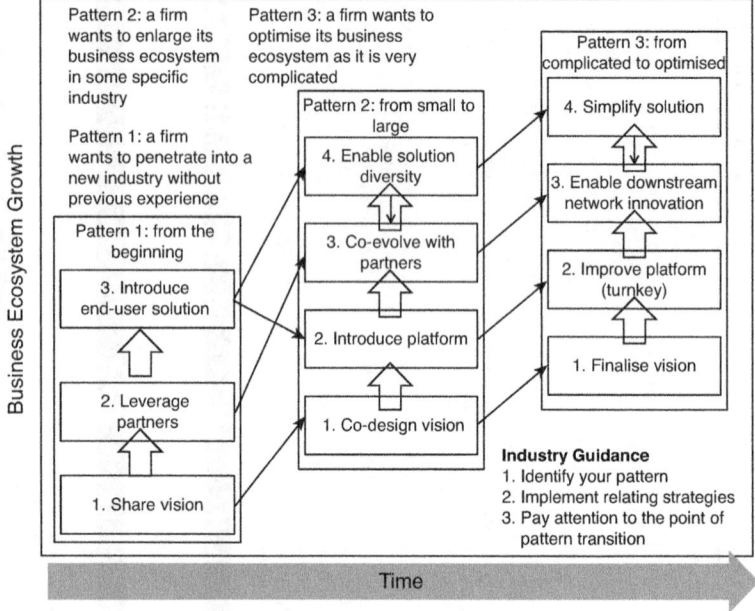

Figure 11.7　Three patterns of nurturing process for practical use

For industry guidance, there are three steps to follow: first, a firm has to identify which pattern matches their business ecosystem status; second, the firm should follow the steps and reach the transition points; thirdly, when the firm reaches the point of pattern transition, the firm has to adopt the new pattern's strategies.

11.5　Discussion on the nurturing process

Literature review suggests that the process of building a business ecosystem has not been explored clearly regarding an emerging industry. This chapter offers a process of how an ecosystem is created from the firm perspective under the context of BELC phases. The key phases and activities are distinguished. Previously, the Nurturing process was basically proposed by Moore as birth, authority, expansion and renewal (Moore 1996). However, in terms of the new and dynamic industry environment, recent literature has added more content to those phases. In terms of the first phase, Moore suggested the creation of a value chain, while this study advises firms to consider two more aspects: succeeding from the previous ecosystem and sharing the product vision for future development, rather than only creating value chains. In the second phase, Moore suggested scaling up the market by introducing market standards. However, in this phase the market is still in an uncertain situation. As a result, in order to meet uncertain issues, more diversified ideas are encouraged instead of the expansion from Moore's view. Furthermore, future product vision should also be co-designed by partners. In contrast, Moore suggested only expanding the market volume. In the third phase, Moore proposed they should be authorities to maintain the advantage, while this study divides the third phase into two phases: Converging and Consolidating. Converging aims to select a typical solution and to select partners grouped around the finalised future product vision. Consolidating is to approach the best solution and integrate the partners' network in order to achieve mass production, which normally lasts for a long period until the emergence of a niche idea. In the Renewing phase, Moore suggested maintaining the barriers to entry as well as setting a high customer switching cost to maintain the partners' network, then to use a new idea to renew the ecosystem. However, in this study, the Renewing phase aims to insert a niche idea, reorganise the partners' network, and cut down the entry barriers in order to renew the original industry into a brand-new industry. In conclusion, the nurturing process developed in this study is more of a co-evolutionary style while Moore's work relies on a competitive style.

11.6 The key constructive elements during the nurturing process

Learning from the nurturing process, four out of the eight constructive elements in blue in Figure 11.8 are highlighted as the key elements in the nurturing process: new vision development, solution platform, functional role and adaptive solution.

The element of 'new vision development' is an informal kind of authority which is proposed by a focal firm to encourage the ecosystem partners (Gulati et al. 2012; Reid & Brentani 2012). Then, the focal firm will share the vision and provide the way to transfer the vision to solution. As a result, the focal firm would introduce the platform for the end-user product's development. The focal firm would learn the lesson from the Emerging phase that their product is not an end-user product but could be the product platform for end-user product development. Thereafter, in the following phases, they start to introduce, co-design

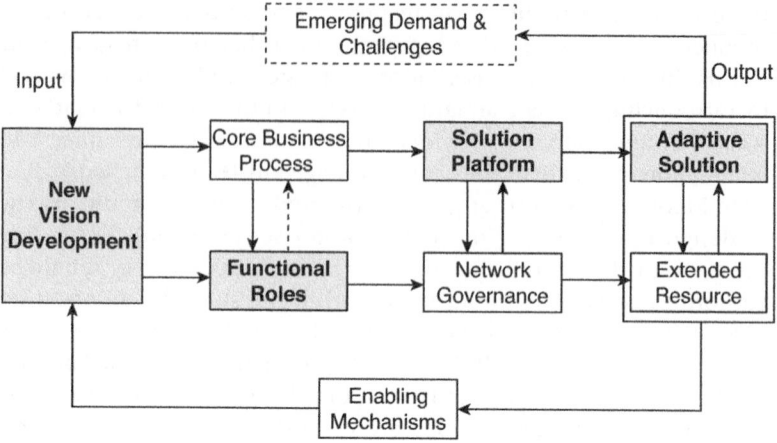

Figure 11.8   Key elements during the nurturing process

and consolidate the platforms and encourage the different partners with various functions to contribute. The platform is the place that facilitates the process of realising the vision. Every partner could contribute to this process. As a result, the end-user solutions will meet the features of those sequential phases and become very adaptive.

The other four elements – core business process, network governance, extended resources and enabling mechanism – will take supportive roles to facilitate other elements' evolution. The key elements will make the focal firm and other partners more easily capture the nature of a business ecosystem and nurture the ecosystem more directly and effectively.

11.7   Conclusion

This chapter developed firms' nurturing process along the BELC.

First, it reviewed three main firms and conducted a cross-projects analysis in order to develop three unique nurturing processes along the BELC.

Second, by cross-firm analysis, this chapter also developed a general nurturing process along the BELC: Emerging phase indicates that firms would share vision, introduce solutions and leverage partners. In the Diversifying phase, firms aim to co-design vision, introduce platform and encourage partners, which boosts solutions variety. The Converging phase demonstrates that firms are to finalise visions, co-design platform, select typical solutions and relevant partners. In the Consolidating phase, firms finalise the products and platform in order to integrate partners to achieve mass production. The Renewing phase is where firms insert new ideas and prepare for the next cycle of the renewed business ecosystem.

Third, from the practical perspective, this chapter also developed three patterns of nurturing process in terms of different industrial situations: from the beginning, from small to large and from complicated to optimised.

At the end, this chapter conducted a discussion on the nurturing process by comparison with Moore's work as well as summarising the key constructive elements along the nurturing process.

12
Conclusion and Discussion

This chapter concludes the book by summarising the research findings, discussing their theoretical and practical implications, and suggesting future research directions as below:

- It reviews the research question and specific research objectives
- It summarises four main research findings including the life cycle of business ecosystems, BE constructive elements, BE configuration patterns, and BE nurturing process from company perspective, as well as the research approaches for exploring BE that is a very complex and dynamic system
- The implications to both theories and practices will be delivered based on those research findings
- Research limitations and the future research into business ecosystem are discussed in order to continue the research in an appropriate direction.

12.1 Research findings

12.1.1 Reviews of the research question and its specific objectives

There are two outstanding industrial issues of uncertainty and interoperability significantly challenging the relevant existing literature. As a potential research area, the business ecosystem has addressed those phenomena. The business ecosystem study aimed to understand the economic community as a whole by highlighting co-evolution activities among different levels of organisations. Business ecosystem thinking had already been applied to industry development in the last decade, however deep understanding of business ecosystem theory is still lacking. As a result, this study aims to enrich the theory of the business ecosystem. Specifically, in order to fulfil the research on the business ecosystem, the key research question was set as '**How to nurture a business ecosystem from firm perspectives?**' which consists of four detailed research objectives to:

- Understand and characterise the life cycle of a business ecosystem
- Deconstruct business ecosystem to identify main constructs
- Identify key configurations of business ecosystems
- Develop a nurturing process for a business ecosystem along the life cycle

12.1.2   Understanding and definition of business ecosystems

The working definition of business ecosystem was first proposed in Chapter 3. Based on the industrial case studies, the proposed working definitions can still hold its position, which recognises **business ecosystem is a community** consisting of differently **interdependent organisations** and stimulating **co-evolution** between partners and their business environment. It is clear that a business ecosystem has its static and dynamic characteristics.

The concept of 'business ecosystem' actually includes two levels of meaning towards business – general business and specific business. In terms of 'general business', the business ecosystems can be distinguished from other types of ecosystems such as natural ecosystems, political ecosystems or innovation ecosystems. The business ecosystems are dedicated to the business development world or value creation, and form the context of commercialisation and transform the value chain. In terms of 'specific business', under the context of 'general business', firms start to coordinate different partners and capture resources in order to deliver particular products or services. Therefore, there are many types of specific business ecosystems, such as electric vehicle ecosystem, 3D printing ecosystem, mobile computing ecosystem and robotics ecosystem. Any business or firm must have its own unique ecosystem.

By deeply exploring the definitions, business ecosystem has many outstanding features: it is composed of interdependent organisations; business opportunities emerge inside the ecosystem to trigger interactions among organisations; organisations symbiose with others of the same mindset; firms inside the ecosystem co-evolve with one another and shape the industry future together.

With the above features, a business ecosystem can generate some outstanding results: community partners generate co-evolution with one another so as to cope with an uncertain future and to sustain industry development. Also, the organisations can rapidly coordinate to commercialise new ideas and thus renew the existing industry.

In order to understand the business ecosystem more clearly, the following four attributes of business ecosystem have been explored by the book: its life cycle, constructs, configuration patterns and its nurturing process, which help us deeply understand its behaviours and implications.

12.1.3   Business ecosystem life cycle

The business ecosystem life cycle (BELC) consists of five phases: Emerging, Diversifying, Converging, Consolidating and Renewing.

- Emerging: new solution is proposed for the emerging market, and then the central firm leverages partners to initiate the supply chain for the new market. This phase ends with features such as a new solution for an emerging market and simple supply chain establishment.
- Diversifying: solution is very diversified to meet the uncertain market. Partners are encouraged to contribute, and partners' network is very flexible with high interoperability. This phase ends with features such as solution diversity and a high level of partners' network interaction.
- Converging: the market is becoming specialised and the selection of solutions has also converged. As a result, the partners' network integrates them for specialised markets. This phase ends with market specialisation, solution convergence and a converged partner network.
- Consolidating: lasts for a long time as the dominant design is approached. The partners' network is stable and forming a close alliance for mass production of that dominant design. This phase ends with such features as dominant design and stable partner network structure.
- Renewing: a niche market is emerging and the partners' network starts to reorganise itself in order to renew the original market. The original market can be replaced by the emerging market or stay for another long period. If the original market is replaced by the emerging market, firms will re-enter phase 1 to experience the five phases again. So this phase ends with features such as a niche market and partners' network reorganisation.

These sequential phases form the BELC. Each phase ends with different statuses, which helps firms to check with the ecosystem development phases. In different phases, firms implement appropriately different strategies.

12.1.4 Constructive elements of a business ecosystem

Constructive elements are founded on the basis of the classical model 'structure and infrastructure' (Hayes & Wheelwright 1984) in order to keep close relationship with classical theories. Structural factors demonstrate long-term-oriented constructive elements which are stable and fundamental, while infrastructural factors are short-term-oriented constructive elements which are more changeable and dynamic. As a result, in each phase of a nurturing phase, relevant constructive elements will be addressed differently.

On the structural side, there are functional roles, solution platform, adaptive solution and extended resources. Functional roles place the players from different organisations (industrial players, quasi-governmental organisation, government organisations) into three types: initiators, specialists and adopters. This book mostly studies the central firms (initiators) who are willing to set up their business ecosystems. An adaptive solution is the key product solution which is initiated by the central firm and contributed to by its partners in order to meet different requirements along the BELC. A solution platform enables partners to easily interact with one another, and provides more niche opportunities to add value. Extended resources are often provided by specialists who are the supporting parts of the business ecosystem operation.

On the infrastructural side, network governance is that which organises the network and coordinates the relationships in order to meet the requirements from partners. New vision development is that which make the players co-design the direction of the industry's development. Core business process indicates that business and operation processes vary according to the BELC in order to implement new product development. Enabling mechanisms development contains market issues, regulation and societal issues.

In all, this finding extends the construct study of 'structure–infrastructure' framework from the firm level, intra-firm level, inter-firm level towards the business ecosystem level.

12.1.5 Configuration patterns of a business ecosystem

- The key elements to identify pattern

A configuration pattern shows how a particular business ecosystem works by fixing and displaying its fundamental elements. From both industrial concerns and literature reviews, two key elements have been identified to categorise patterns of configuration: Solution platform openness and Solution diversity. The platform indicates an interface for players' interaction and participation. The Solution consists of partners' contributions which are developed based on that Solution platform.

- The configuration pattern of a business ecosystem

There are seven patterns relating to Solution Platform Openness and Solution Diversity shown in Figure 12.1 Pattern 1 – simple solution ecosystem; Pattern 2 – Platform enabling ecosystem; Pattern 3 – Platform integrated ecosystem; Pattern 4 – Platform coordinating ecosystem; Pattern 5 – Platform co-evolving ecosystem; Pattern 6 – Open community ecosystem coordinated by few companies, and Pattern 7 – Open community ecosystem.

Besides the seven patterns described, there are no cases found to demonstrate the other two potential patterns. Actually they are not logical, because if the business ecosystem has a very open or less open platform, the solution cannot at the same time be closed.

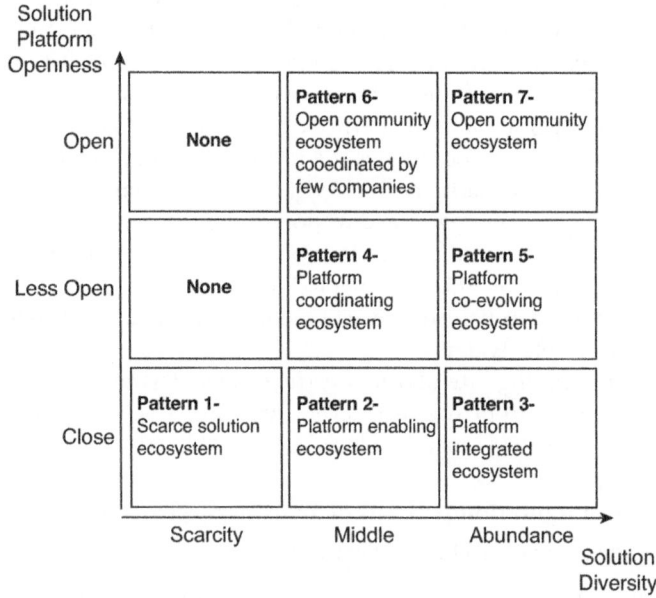

Figure 12.1   Conclusion about the configuration pattern

- The pattern evolution model along nurturing process

Besides static pattern identification, the evolution model pattern along the business ecosystem nurturing process is also highlighted in this study. There are many evolutionary path models. Three paths are identified based on the three main case studies. ARM adopted a very open path, and Intel began to open the door to its partners. MTK led an opportunity-driven pattern. By comparing these three typical evolutionary pattern models, a general evolutionary pattern model is delivered in two different contexts: in the context of established industry, a firm would come closer to pursuing industry efficiency by using the model Pattern 1-3-2-2-2. In the context of an emerging industry, firms are more open to collaboration and co-evolution with one another, so they use the model of Pattern (4/2)-5-4-4-(4/2).

12.1.6 Nurturing process along life cycle

Regarding different features of each life-cycle phase, firms nurture the business ecosystem as follows: in Phase 1, Emerging, firms start to introduce a new solution to persuade partners to work together. As a result, the supply chain is initiated for this new solution. In Phase 2, Diversifying, firms introduce the product platform and then encourage partners to co-design the end-user solutions. As a result, a diversified solution is introduced and many partners interact to deliver those solutions. In Phase 3, Converging, central firms further improve the solution platform with partners and enable them to start to select some typical solutions which meet the market requirements; as a result, partners will be selected and the market becomes specialised. In Phase 4, Consolidating, firms continue to improve the solution platform in order to finalise the design and maintain competitive advantage. So the central firm will integrate partners in order to improve industry efficiency and achieve mass production. In Phase 5, Emerging, firms will introduce some niche ideas in order to persuade partners to enter another relevant emerging industry. The partners' network will be reorganised in terms of the niche ideas. Then the business ecosystem again enters Phase 1, leveraging partners to rapidly commercialise the new ideas. These are the five phases of the BELC, which also sustain the business ecosystem development.

Besides the nurturing process along the BELC, there are also three patterns of nurturing process regarding the business ecosystem scope: Pattern 1 – If a firm wants to penetrate into a new industry without previous experience; Pattern 2 – If a firm wants to enlarge its business ecosystem in some specific industry; Pattern 3 – If a firm wants to optimise its business ecosystem as it is very complicated.

12.1.7   Integration of four research findings

Those four research findings are integrated into Figure 12.2 in order to demonstrate their internal relationships.

In terms of the life-cycle study, five phases emerge first from one real main case and are then refined by the other two main cases. The BELC demonstrates itself as having five sequential phases: Emerging, Diversifying, Converging, Consolidating and Renewing. In order to identify the boundary of each phase, there are also six phase statuses: Pre-emerging, Post-emerging, Post-diversifying, Post-converging, Post-consolidating and Post-renewing.

Following the five phases of the life cycle, the business ecosystem sets the context for the central firm. In each phase, the central firm would nurture their own business ecosystem with different pathways. These nurturing processes are developed by comparing the cases. Key common activities are selected as the nurturing steps.

As a construct is static, it was derived from the six static phase statuses. The structure and infrastructure model was selected to deliver the relevant constructive elements. Each constructive element has a different impact on each phase. After cross-phase analysis, the generalised constructive elements were finally identified.

As to the study on the configuration pattern variables, the key identical dimensions should not only be derived from constructive elements, but also match the key industry issues of uncertainty and interoperability. So the most representative and critical elements were selected to distinguish patterns. Then those patterns would also form two typical evolutionary models with two different industrial contexts along the BELC.

12.1.8   Business ecosystem exploration methodology

The business ecosystem research has attracted great attention in academic and industrial communities. More research will be conducted in its arena which is very complex and dynamic. This book introduces our methodology following the research question we want to explore. Our case studies are not based on the companies but within their new business development projects. The project evolutionary mapping and storytelling fundamentally inspire us to understand the business ecosystem's real meanings – its core, boundary, features, change, and all kinds of connectivities and interactions.

12.2 Implication to theory

This research contributes to the theory of the business ecosystem in terms of its life cycle, nurturing process from the firm perspective, constructs and configuration pattern as shown in Table 12.1 by evolution from previous network theories.

- In previous literature of the business ecosystem, Moore proposed only a four-phase BELC. This study expands Moore's BELC by giving further detail of the phases context, expanding them into five phases, as well as enriching the content of each phase and highlighting different priorities. This life-cycle study is also extended from PLC, TLC, firm growth and ILC studies.
- There is lack of studies on how firms nurture the business ecosystem in previous literature. This study enriches, from the firm's perspective, the detailed process of nurturing its business ecosystem. This study also explores in detail the three types of nurturing process, with different starting points.
- In previous literature, different types of business ecosystem were studied, however, detailed fundamental research as to the constructive elements of a business ecosystem had not yet been conducted. This study provides constructive elements of the business ecosystem from the structure and infrastructure perspective, which also extends the study of constructs of the firm, intra-firm, inter-firm and internationalised inter-firm manufacturing systems.
- This study also addresses the configuration of the business ecosystem area in order to get an integrated understanding, which has not been explored in previous business ecosystem literature. Seven different patterns have been identified. A general configuration pattern selection model is also identified along the BELC. This study not only explores the configuration patterns at the business ecosystem level, but also reflects the studies on the international manufacturing system and global supply chain levels.

In summary, the findings in this study about the business ecosystem are developed from manufacturing system research and its related literature.

12.3 Implication to research methodology

- Network study framework

A multiple-perspective research framework is presented in detail to review the network theories in a comprehensive way: context, life cycle,

Scale: Market

Phase 1	Phase 2	Phase 3	Phase 4	Phase 5
Emerging	Diversifying	Converging	Consolidating	Renewing

Share vision → Introduce solution → Co-design vision → Enable solution diversity → Finalise vision → Select Solution → Consolidate platform → Improve & Finalise Solution → Initiate & Capture niche idea

Leverage Partners → Introduce platform → Encourage Partners → Co-design platform → Select Partners → Integrate Partners → Reorganise partners → Start new life cycle

Decline

Life cycle & Phase Status ⇒ Nurturing Process

Status 0 | Status 1 | Status 2 | Status 3 | Status 4 | Status 5

Continued

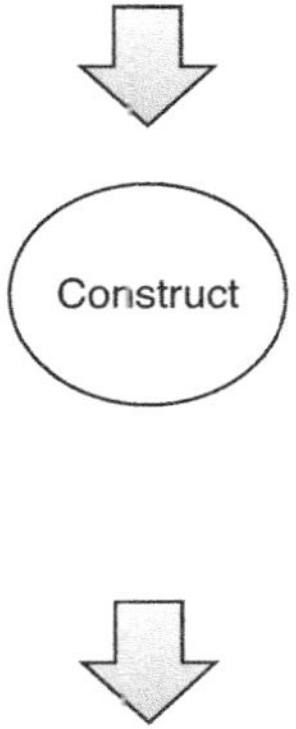

Constructs	Status 0: Pre-Emerging	Status 1: Post-Emerging	Status 2: Post-Diversifying	Status 3: Post-Converging	Status 4: Post-Consolidating	Status 5: Post-Renewing	Constructive elements
Structure	niche idea	emerging solution	diversified solution	selected solution	product as dominant design	niche idea	Adaptive Solution
	separated roles:	role (centric firms with value chain initiated)	roles (centric roles with crossed value chain)	roles (centric firm with selected value chain)	roles (centric firm with integrated partners)	roles (centric role to reorganize partners)	Functional Roles
			platform with easy access	Integrated product platform	continuing improvement of product platform		Solution Platform
		extended existing resource	various resource combination	selected resources	Integrated resource		Extended Resources
Infrastructure	new vision initialled	common vision initiated	vision co-designed	finalised vision		new vision introduced	New Vision Development
		leveraged relationship	diversified relationship	selected relationship	close relationship		Network governance
		platform based simple value chain	platform based flexible network	platform based converging network	close network around platform		Core Business Process
		emerging market	market diversity and flexible societal and policy system	market specialised and adaptable societal and policy system	supportive societal and mature policy system, stable market	market transition	Enabling Mechanism Development: Market, societal & policy system

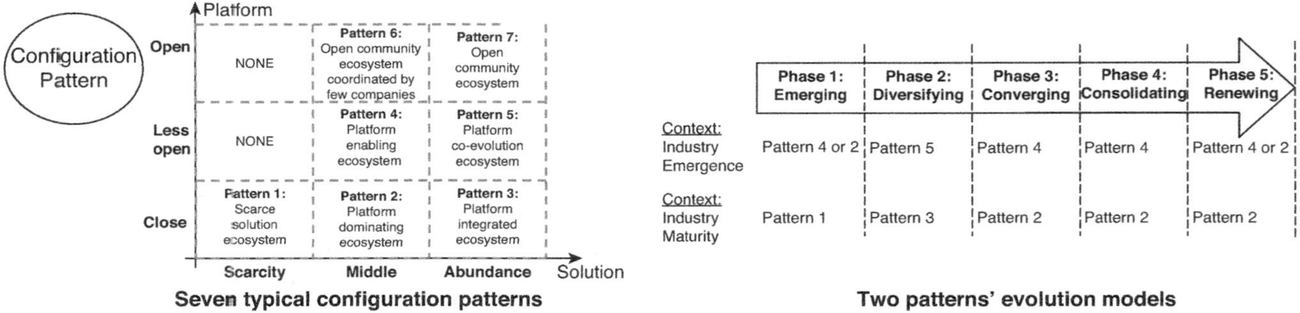

Platform

	Scarcity	Middle	Abundance
Open	NONE	Pattern 6: Open community ecosystem coordinated by few companies	Pattern 7: Open community community ecosystem
Less open	NONE	Pattern 4: Platform enabling ecosystem	Pattern 5: Platform co-evolution ecosystem
Close	Pattern 1: Scarce solution ecosystem	Pattern 2: Platform dominating ecosystem	Pattern 3: Platform integrated ecosystem

Solution

Seven typical configuration patterns

	Phase 1: Emerging	Phase 2: Diversifying	Phase 3: Converging	Phase 4: Consolidating	Phase 5: Renewing
Context: Industry Emergence	Pattern 4 or 2	Pattern 5	Pattern 4	Pattern 4	Pattern 4 or 2
Context: Industry Maturity	Pattern 1	Pattern 2	Pattern 2	Pattern 2	Pattern 2

Two patterns' evolution models

Construct

Figure 12.2 Research findings integration

nurturing process constructs and configuration pattern. Context is the driving force for this study; life cycle is to study the evolution of a business ecosystem, and the nurturing process is to describe the process of a firm's nurturing its business ecosystem along life-cycle phases. Constructs present the fundamental factors used to construct a business ecosystem and configuration patterns indicate a synthesised model of firms' strategy within their business ecosystems.

- Five-step data analysis method

A synthesised data analysis method is proposed with five steps: 1) data mapping, 2) intra-firm cross-case analysis, 3) inter-firm cross-case analysis, 4) cross-phase generalisation and 5) building theory.

The core of this method has two aspects: the first is to conduct intra-firm/inter-firm cross-case analysis. The draft findings are concluded from real data. Then other data are continuously used to improve the draft findings until the finding content reaches a stable status. The second is to conduct cross-case analysis along the life-cycle phases. This method generalised the result along time.

- Configuration pattern study framework

There are three steps in studying configuration patterns: 1) to select the most representative dimensions to identify configuration patterns which are constructive elements to represent specific purposes. For example, this study chooses constructive elements to reflect the two outstanding industrial challenges; 2) to identify typical patterns from real cases by using the key elements. The pattern variations demonstrate the variety of ecosystem strategies; 3) to study pattern selection along the BELC. This part aims to demonstrate the different potential pathways for nurturing a business ecosystem.

12.4   Implication to practice

As seen from Table 12.1, knowledge of the life cycle of a business ecosystem would help firms to realise the current context of their business ecosystem. In each phase of the life cycle, there are also some typical activities.

Regarding the nurturing process, firms can learn a general way to build up a business ecosystem along the five sequential phases by following different steps.

Constructing a business ecosystem helps firms to understand that fundamental factors have big impact on business ecosystem development. Also, in each phase of the nurturing process, different constructive

Table 12.1 Theoretical and practical implication

Research Findings	Items in detail	Theoretical implication	Practical Implication
The relationship between existing theory and business ecosystem	• **Research connection: life cycle; nurturing process; constructs; configuration pattern**	• Using four research focus to connect manufacturing system from level of firm, intra-firm, inter-firm towards internationalised inter-firm	/
Life cycle	• **Five sequential phases:** Emerging; Diversifying; Converging; Consolidating; Renewing	• Enrich product life cycle, firm growth life cycle, industry life cycle, technology life cycle	• Identifying the business ecosystem in specific phase of life cycle and understanding the typical activities in each phase
Constructive elements	• **Structural side:** functional roles, adaptive solution, solution platform, extended resources **Infrastructural side:** network governance, new vision development; core business process; enabling mechanism development	• Constructing a business ecosystem skeleton • Connecting a business ecosystem with classical theory	• Providing a way to investigate different fundamental factors of firm's business ecosystem • The elements' impact on each phase's nurturing process and firm strategy
Configuration pattern	• Identified by two dimensions: **solution diversity and platform openness: seven patterns** • Pattern evolution along life cycle	• Building up a business ecosystem configuration pattern theory • Configuration pattern evolution implication to ecosystem growth	• Providing a synthesised model to present firm strategy in a business ecosystem • Practical guideline for a business ecosystem's strategy and growth

Continued

Table 12.1 Continued

Research Findings	Items in detail	Theoretical implication	Practical Implication
Nurturing process	• **Detail nurturing process in each phase of life cycle** • Three situations/patterns of nurturing process • Key constructive elements during the nurturing process	• Describing the ecosystem growing process • Enriching Moore's theory • Capture the key elements	• Generalising the way to build up the business ecosystem • To find the right way to renew the business ecosystem
Research methodologies	• **Network study framework:** context, life cycle, constructs, nurturing process, configuration • **Five-step data analysis methods** • **Configuration Pattern study framework:** pattern identification and evolution model study	• The comprehensive framework to lead network study • Research findings are learned from real data and improved by real data • The brief way to categorise the different networks and their evolutionary model	/

elements play different and vital roles. As a result, firms can learn from these constructive elements how to build up a business ecosystem.

The configuration pattern is a synthesised model to demonstrate firms' strategy in nurturing the business ecosystem. By going through the life cycle, the adoption of different configuration patterns would result in different consequences. So, firms can learn to adopt different configuration patterns as their key strategy to achieve different purposes.

Besides these four perspectives, some conclusions can also be made about different industrial contexts: emerging industry and established industry.

- To emerging industry

An industry emerges from and with uncertainty and dynamic change. Business ecosystem thinking can help deal with uncertainty and help to create a platform to encourage partners' co-creativity. Also the five phases (Emerging, Diversifying, Converging, Consolidating and Renewing) can guide the firm's practice to build a new ecosystem.

- To established industry

In a less established industry, ecosystem thinking would help firms identify their business environment and coordinate existing resources in an appropriate way. This would make firms' business ecosystems stronger and more mature.

In a well-established industry, ecosystem thinking would enable firms to reconsider the constructive elements of the business ecosystem in order to find niche ideas to renew the current industry and make the ecosystem sustainable. This process would enable mature industry to continue growing with a substitute product instead of declining.

12.5   Research limitation

There are also some limitations for this research, although a structurally designed research methodology has been followed. However, limitations as shown in Figure 12.3 should help readers comprehensively understand this book.

12.5.1   Case studies selection scope and phase identification

The case studies are central to identifying how a firm nurtures its business ecosystem. The process view is adopted and sequential projects are selected in order to identify their nurturing process. The number and content of projects matter in creating the process description, so this book tried to find as many projects as possible that had unique contributions to their nurturing process, in order to comprehensively identify the process. So far, seven exploratory cases and three main case studies, with three sub-cases each, have been studied.

However, as learned from those cases, it is also a challenge to identify phases to a very precise degree. All of these five phases are identified by learning from real cases and refined by real cases.

Besides this, as this study aims to explore a firm's business ecosystem, it is also a challenge to collect data in a very comprehensive way. In this study, typical organisations and key projects were selected in order to demonstrate a firm's nurturing the business ecosystem along the life cycle.

12.5.2   The relevant literature scope

Seven relevant theoretical domains were reviewed including supply chain management, business network, strategic alliance, industrial organisation, industrial cluster, global manufacturing virtual networks (GMVNs), and open innovation, in order to identify key theoretical gaps and business development trends. However, whether there are still other relevant theories that should be reviewed remains a question. So this book proposed a framework which it used to organise the theories into four streams: firm-based, intra-firm, inter-firm and internationalised inter-firm in order to make the literature review more feasible and comprehensive.

12.5.3   Theory generalisability

The mobile computing industry, studied in this book, has some outstanding features. It is fast-moving, highly innovative, always changing and very uncertain. Furthermore, strict case selection criteria were developed in order to make the research findings as generalised as possible. As a result, the research findings could be properly applied to other industries with those features.

However, due to the different natures of the evolution of different technologies, different firms and the evolutionary stages of industry development, the generalisability of this study's findings may also be challenged.

In terms of the evolution of technology, technologies such as chip platforms become more open than before, which re-configures the patterns of partners' interaction.

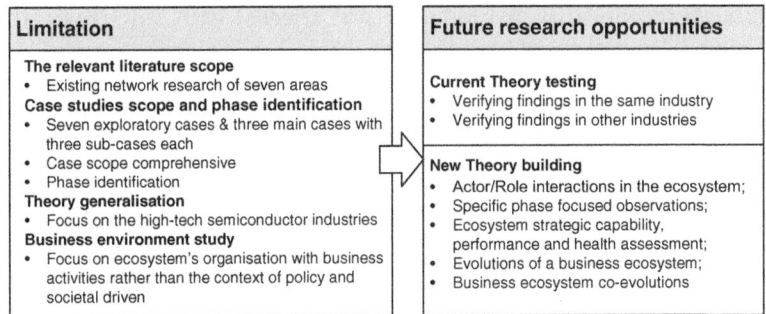

Figure 12.3 Research limitation and future research opportunities

In terms of different firms, they have different historical pathways of development and accumulate unique features.

In terms of evolutionary stages of industry development, different stages have different contexts which can have great impact on firms' strategies and external dynamic environment.

12.5.4 Business environment study

Generally speaking, the business ecosystem consists of the various levels of organisation and relevant activities (Moore 1996). However, many works on the study of business ecosystems are focused more on the system itself rather than its business environment issues like policy and societal systems. This research studies industrial phenomena at a system level and their impact on an ecosystem organisation's strategies. However, it also lacks understanding of the full meaning of contexts such as the market, policy and societal influence.

12.6 Future research opportunities

Regarding the research limitations, there are also two areas for future research shown in Figure 12.3: theory building and theory testing. Theory building aims to build ways to uncover parts of the business ecosystem from depth study and width study. Theory testing is to demonstrate this book's findings in the same industries or other industries.

12.6.1 Current findings testing

So far, life cycle, nurturing process, construct and configuration pattern are all identified in this book. Regarding research limitations, even though the research methodology was selected with strict rules, the generalisation of those findings may also face challenges. As a result, all these four research findings should be tested in the same industries as well as in other industries which would enable the research findings to be more feasible and tangible. Moreover, after testing in different industries, industry experience will also help refine findings about the business ecosystem.

12.6.2 Future theory building

There are also several areas for further exploration.

- **Types of actors and role-focused research**: as this study focuses more on the keystone type of firm, more kinds of organisations and actors should be added to the business ecosystem framework, such as government, industrial associations and other relevant organisations who contribute to ecosystem development. The new study can be conducted from the perspective of different organisations and their activities. After that, more comprehensive understanding, especially the interactive relationships among the actors, can be achieved on the BELC from the beginning.
- **Phase-based research**: more detailed study should be conducted in order to explore the different strategies and activities based on each phase of the BELC.
- **Business ecosystem's strategic capability and health auditing**: capability is a combination between resources and business process and determined by the configurations. It is essential to understand the key dimensions of ecosystem capabilities and their relationship with configurations. Furthermore, performance and health auditing work (evaluation) should be developed to understand the ecosystem and the space for future improvement.
- **Business ecosystem cooperation and co-evolution**: besides the nurturing business ecosystem from firm perspectives, more detailed study should be addressed to business ecosystem operational mechanisms – how different firms interact in each life-cycle phase and how they renew business ecosystems.

The findings of this research and existing knowledge have already provided a good foundation for future investigation and research. Business ecosystem not only provides a metaphor but actually opens a new gate demonstrating a new pathway and methodology for a sustainable business development.

Appendix

Table 1   Questions' list

Focal firm corporate level	Interviewee
Could you brief your company history please?	Director/CEO
Could you give an introduction about your firm's series of products?	Director/CEO
Please describe your partners and suppliers/customers and your company's position in that partners' ecosystem.	Director/CEO
Please introduce some typical projects along the company's evolution? Why do those projects matter in your company's growth? What are the key issues in the emerging industries?	Director/CEO
What have you learned from those projects and how do they relate to your future business strategies since you have nurtured your business ecosystems?	Director/CEO
Then, in terms of each project, the following interview questions will be asked of the managers or directors from focal firms and business ecosystem partners.	

Focal firm's project level	
Why this project? What are the key issues to penetrate into emerging industries and promote uncertain products?	Manager
In terms of each project, please explain the project's plan and activities as well as the team members.	Manager
How to incubate new ideas for industry future and deliver the solutions?	Manager
What are the key projects' tasks? Such as make-or-buy decision, partners' identification and selection? How to drive your partners and quickly set up a supply chain?	Manager
How do you learn from the project? Do you regard it as a step to nurture your firm's business ecosystem?	Manager
What are the key factors during the nurturing process?	Manager
What are the other methods to nurture the business ecosystem, such as financial investment, policies and social norms?	Manager
After your project, how to keep the relationship with your key partners? Are there any sustainable and systemic methods?	Manager

Continued

Table 1   Continued

Business ecosystem partners' level	
Why does your company participate in this project?	Director/ Manager
How does your company set up the relationship with the focal firms?	Director/ Manager
Could you describe the process of specific project and your task?	Director/ Manager
How does your company get benefit from this collaboration?	Director/ Manager
Have you realised your collaboration actually co-nurtures a business ecosystem with the focal firm and other partners?	Director/ Manager

	Phase 1: Emerging	Phase 2: Diversifying	Phase 3: Converging	Phase 4: Consolidating	Phase 5: Renewing
1a Mobile	1). In 1994, ARM developed IP (ARM7) with Low power and low size. 2). They aimed to persuade TI to adopt their IPs, but without success. At that time, TI was looking for low-power processors for mobile phone development. 3). ARM talked with Nokia w\no was also aiming to develop low power IPs. 4). With Nokia's help, TI began to adopt ARM's IP. 5). The first supply chain was built up and the first mobile phone Nokia 6110 was delivered in 1997. 6). ARM realised it was very important to leverage partners and get feedback from OEM in order to create new supply chain.	1). After TI firstly adopted ARM's IP, they two started to co-design the future mobile solution. 2). ARM designed the IPs with TI and Nokia's feedback. 3). In order to highlight the ARM performance, ARM also supported software partners to develop operating system and relevant application software to enhance ARM's IP performance. 4). ARM also shared IP with design tools partners, manufacturing partners in order to reduce product lead time.	1). After ARM firstly launched mobile phone with Nokia, other IC design firms and OEMs began to adopt ARM's IPs. 2). ARM's Low power solutions were winning the competitive advantage. 3). ARM built up the strong alliance with TI and Nokia	1). ARM7 was not good enough to meet the market requirement. 2). ARM9 was improved in processing speed, low power and size as well. 3). ARM continue to cooperate with TI and Nokia in ARM9 4). ARM also supported OS partners and ISV with relevant development tools.	1). ARM's IP could be used not only in mobile phone market, but also in other embedded market. 2). ARM7 was re-designed for the embedded market & ARM began to find new partners for the IP
1b Leader Partner strategy		1). The embedded market is huge and very fragmented, and no one was dominating in this market. 2). Market requirement research with four top priorities, which determined 32bit would be first choice. 3). ARM recycled ARM7 and developed it into Cortex M3, as a result, ARM also succeed from ARM7's ecosystem. 4). ARM also co-marketing with leader partners-Luminary Micro 5). ARM also build up connected community to support other non-direct business partners to add value to ARM's products for example supporting OS partners and third party software	1). the embedded market specialised, ARM and Luminary Micro provided Stellaris family processor. 2). ARM and lead partner formed small alliance for new IP promotion 3). They also succeeded strong software ecosystem from ARM7.	1). More partners joining as the leader partners. ST was also selected in order to expand ARM's IP market share in the embedded market. 2). ARM implement the complete process of leader partner strategy: To identify leader partners as top 5 player or player with strong innovation capability and similar strategies; then, ARM got Requirement from leader partners: for example, With high performance requirement; response for interrupt; low coding density and combination the two types of coding system. So far, More than 100 customer chose SMT32 (ARM based)	
1c IP category			1). As the embedded market was very specialised, three markets have been highlighted as mobile computing, industry control and low-level embedded system. 2). Regarding to these three market, ARM's existing IPs were developed with performance priority. 3). So in order to meet different market, ARM's IP were then customised into Cortex A, R, M series IPs.	1). ARM continued to implement leader partners strategy in terms of these three streams. 2). Those IPs were improved all the time and aiming to approach the best solution 3). The previous business ecosystem partners are also reorganised regarding to those three streams.	

Figure 1 Data from ARM cases

	Phase 1: Emerging	Phase 2: Diversifying	Phase 3: Converging	Phase 4: Consolidating	Phase 5: Renewing
2a PC	1). In early stage of PC industry, IBM and Apple were the two dominating players who were with different organisational structure. 2). Apple was very vertical integration and IBM was coordinating a partner network. 3). Intel as a processor provider, was one of very important suppliers of IBM's network.	1). IBM's model won the competitive advantage against Apple's model. So many players began to imitate IBM's model to coordinate different players who use IBM's PC design to produce PC. 2). Intel as a key supplier, was supporting different industry architecture from various OEMs. 3). Those OEMs used Intel's processor to differentiate their end-user products	1). Intel discovered it was very difficult to support so many partners. 2). IBM set up a IP fee to charge players who use IBM's PC design. 3). IBM's architecture design was an obstacle for Intel's development of processor. As IBM architecture cannot make full use of Intel's processor. 4). Intel began to consider about platform based collaboration leader platform strategy and finally set up PCI standard, which easily connected with many peripherals processors. 5). PCI enabled Intel develop their processor without restriction. Intel's processors were compatible to other peripheral chips by generations.	1). In order to maintain competitive advantage, Intel continue to introduce standards: USB, Wimax to develop more advanced technology based processors. 2). Intel followed Moore's law to develop more advanced technology based processors. 3). Intel set up stable partners' network, for example, the close alliance with Microsoft to maintain the barrier	1). The PC industry was maturated, Intel began to consider about entering mobile phone market. 2). Intel got the license StrongARM from a lawsuit settlement with DEC 3). Intel decided to enter mobile and encourage partners to enter new market
2b Xscale			1). As Intel was very successful in PC industry with 'Win-tel' dominating strategy, Intel did not initiate many partners interaction and get rid of diversifying phase. 2). Intel just launched close alliance with Motorola in order to reach the dominant design as in PC industry.	1). More and more partners were adopting ARM's IP to develop mobile phone, so there were too many competitors against Intel. 2). Intel also outsourced manufacturing to TSMC in order to reduce the lead-time as TSMC was familiar to manufacture ARM's IP.	1). Manufacturing and maintenance cost is high as Xscale is not Intel's architecture 2). AMD threatened Intel PC serve processor market 3). Not many support from software side which slow down the supply chain process. 4). Intel sold Xscale to Marvel.
2c Atom	1). Asus provided low version of Notebook and named it as EeePC 2). Intel renamed it as Netbook in order to distinguish it from notebook and maintain notebook's benefit. 3). Intel thought it was good chance to re-enter the mobile market by starting with Netbook. 4). They developed new low power platform-Atom chips.	1). Intel not only initiate the concept of Netbook but also the concept of MID as mobile internet device with OEM (Aigo). 2). Intel also built up Moblin ecosystem based Linux foundation, which encouraged software partners to develop OS and application software based on Atom chips. 3). Intel also provided much financial support to encourage partners to develop complementary products to Atom chips.	1). In order to maintain competitive advantage, Intel also imitated the strategy in PC industry, Intel Introduced Netbook industry standards and persuaded partners with Netbook standard 2). Intel also selected two OEMs as Lenovo and Aigo to develop MID However, Lenovo quit, Aigo's MID did not won good market performance. 3). Intel kept improving its Atom solutions.		

Figure 2 Data from Intel Cases

	Phase 1: Emerging	Phase 2: Diversifying	Phase 3: Converging	Phase 4: Consolidating	Phase 5: Renewing
3a VCD & DVD	1). MTK was spinned off from UMC with strong research capability on multimedia. 2). MTK decided to penetrate into VCD as they have relevant design capability. 3). The VCD market was dominated by USA and Japan firms. 4). MTK was also able to produce single chip solution with low cost. 5). This solution cut down high technology barrier.	1). As MTK was based in Hshinchu science park, so MTK firstly found the customers from those big OEMs within the park. 2). MTK sold the Turnkey model to various partners in Hsinchu park as well as provided the software design kit.	1). Many partners in Hshinchu park were entering mainland China market, So MTK followed them. 2). MTK also enabled local OEMs innovation based on MTK's solution, especially in Shenzhen manufacturing network to produce the VCD.	1). MTK still keep its solution with low cost and low entry barrier 2). MTK continued to improve its solution into the speed of 24X	1). As the DVD market emerged, MTK quickly upgraded VCD solution into DVD solution to renew industry. 2). MTK also succeeded the VCD partners' network
3b Mobile 2.3	1). MTK has its strategy to enter the new market with 70% original technology and 30% new research. 2). The mobile phone market was massive, so they began to enter this market as they have 70% technology of Mobile phone chips comparing with the DVD technology.3 also provided the Turnkey model of mobile phone solution 4). MTK cannot persuade BenQ adopt the solution, but finally collaborated with a small local Taiwan OEM	1). They followed the VCD and DVD strategy and finally enabled shanzhai manufacturing network' innovation, especially those thousands of small entrepreneurs.	1). MTK began to help independent design house partners regarding their performance, which could help MTK develop system design based on MTK's Turnkey solution. 2). MTK also did much trainings for local design engineers.	1). MTK continued to improve the Turnkey model as they added more novel function into the solution for example : MP3. 2). Finally, low cost and luxury mobile phone existed in market. 20% of china mobile phone shipment was achieved by MTK based shanzhi phone in 2008.	1). Currently, more functions were required and embedded into single mobile phone, as a result, the Smartphone market was emerging. 2). MTK introduced low cost of Smartphone solution Succeed from 2G mobile network
3c Smartphone		1). MTK introduced Smartphone succeeding from 2G firm network; 2). MTK collaborated with hardware partners in order to get 3G license for example, WCDMA from Qualcomm, and acquiring ADI to get TD-SCDMA 3). MTK joined Android community to let their solution more compatible to other lead OEMs.4). MTK co-evolved with local software partners: embed QQ into its design. 5). MTK tried to connect Carrier; penetrate into different 3G standard 6). MTK also provided design support and training session;	1). MTK continued to improve the Turnkey model 2). MTK closely design by adopting two 3G license: WCDMA and TD-SCDMA 3). MTK selected partners regarding their performance (IC design, IDH)		

Figure 3 Data from MTK Cases

Notes

2 Industrial Challenges

1. www.arm.com
2. ARM packages ISA and micro-architecture together to deliver different generations of IPs.
3. 'Can iPhone maintain its initiate momentum' Ecch case, ref. 508-117-1, 2008.

5 ARM Nurtures the Business Ecosystem from the Beginning

1. There are some other technical terms equivalent to the word 'IP' in the mobile computing industry such as chips, architecture, processor and platform from different perspectives: 1) as IP is part of Chips, IP is also regarded as the core of chips or chips with basic function; 2) as IP is the integration of instruction set architecture, ARM's IP is also to demonstrate ARM's architecture; 3) because IP has computing function like processing data, ARM's IP is also called ARM's processor; 4) as many IC design partners design their own chips based on ARM's IP, they also called ARM's IP a platform.
2. www.arm.com
3. Normally embedded market is regarded as the electronic market excluding PC industry.
4. ST is short for STMicroelectronics, which is one of the world's largest semiconductor companies with net revenues of US$8.08 billion in 2013. Data from: http://www.st.com/st-web-ui/active/en/about_st/st_company_overview.html
5. http://www.arm.com/products/processors/index.php

6 Intel Re-Enters the Mobile Computing Business Ecosystem

1. www.intel.com
2. From http://www.intel.com/about/companyinfo/museum/exhibits/moore.htm
3. DEC is short for Digital Equipment Corporation, a processor manufacturer from the 1960s to the 1990s (Schein 2003).
4. www.intel.com
5. 'The Jonney Machine' Nam 11.12.07, http://members.forbes.com/global/2007/1112/024a.html
6. Li & Li 2009. Netbooks: learn how you can participate in this exciting category, Intel IDF.
7. 'Netbook Market Forecast & Business Strategy', Koo (2008).http://www.display-bank.com/eng/report/report_show.php?mode=download&id=519&c_id=49

8. http://en.aigo.com/
9. Zhang, Presentation at Intel IDF 2009, 'Developing compelling mobile platforms that deliver the full internet'.
10. http://www.instat.com/products.asp

7 MTK Enhances the Business Ecosystem Efficiency

1. http://www.mediatek.com/en/corporate/index.php
2. 'China handset suppliers prefer MediaTek mobile solutions' http://www.eetasia.com/ART_8800527709_480100_NT_6653a1d8.HTM
3. http://www.three-g.net/3g_standards.html
4. 'China's 3G Network Deployment Update' iSuppli, 2009 http://www.isuppli.com/china-electronics-supply-chain/marketwatch/pages/chinas-3g-network-deployment-update.aspx
5. http://source.android.com/about/philosophy.html
6. http://www.chinamobileltd.com/

References

Adner, R., 2006. Match your innovation strategy to your innovation ecosystem. *Harvard business review*, 84(4), p. 98.

—— 2012. *The wide lens: A new strategy for innovation*, Penguin.com.

Adner, R. & Kapoor, R., 2010. Value creation in innovation ecosystems: How the structure of technological interdependence affects firm performance in new technology generations. *Strategic management journal*, 31(3), pp. 306–333.

Adomavicius, G., Bockstedt, J., Gupta, A., Kauffman, R.J., 2006. Understanding patterns of technology evolution: An ecosystem perspective. In *System Sciences, 2006*. HICSS'06. Proceedings of the 39th Annual Hawaii International Conference on System Sciences, p. 189a.

Agarwal, R., Sarkar, M.B. & Echambadi, R., 2002. The conditioning effect of time on firm survival: An industry life cycle approach. *Academy of management journal*, 45(5), pp. 971–994.

Anggraeni, E., Den Hartigh, E. & Zegveld, M., 2007. Business ecosystem as a perspective for studying the relations between firms and their business networks. In ECCON 2007 Annual Meeting, Bergen aan Zee, The Netherlands.

Astley, W.G. & Fombrun, C.J., 1983. Collective strategy: Social ecology of organizational environments. *The academy of management review*, 8(4), pp. 576–587.

Audretsch, D.B. & Feldman, M.P., 1996. Innovative clusters and the industry life cycle. *Review of industrial organization*, 11(2), pp. 253–273.

Bannerman, P.L. & Zhu, L., 2008. Standardization as a business ecosystem enabler. In G. Feuerlicht & W. Lamersdorf, eds. *Service-oriented computing – ICSOC 2008 workshops*. Lecture notes in computer science. Springer Berlin Heidelberg, pp. 298–303. Available at: http://link.springer.com/chapter/10.1007/978-3-642-01247-1_30

Bell, G.G., 2005. Clusters, networks, and firm innovativeness. *Strategic management journal*, 26(3), pp. 287–295.

Bergman, E.M. & Feser, E.J., 1999a. Industrial and regional clusters: Concepts and comparative applications. *The web book of regional science*. http://www.rri.wvu.edu/WebBook/Bergman-Feser/contents.htm.

—— 1999b. Industry clusters: A methodology and framework for regional development policy in the United States. *Boosting innovation: The cluster approach*, Organisation for Economic Co-Operation and Development, Paris, pp. 243–268.

Blackhurst, J., Wu, T. & O'Grady, P., 2004. Network-based approach to modelling uncertainty in a supply chain. *International journal of production research*, 42(8), pp. 1639–1658.

Booth, J.A., 2008. Consolidating the MCU market around the ARM architecture. *Embedded*. Available at: http://www.embedded.com/electronics-blogs/industry-comment/4026853/Consolidating-the-MCU-market-around-the-ARM-architecture.

Boretos, G.P., 2007. The future of the mobile phone business. *Technological forecasting and social change*, 74(3), pp. 331–340.

Borrus, M., Tyson, L.D. & Zysman, J., 1986. Creating advantage: how government policies shape international trade in the semiconductor industry. *Strategic trade policy and the new international economics*, pp. 91–114.

Bronfenbrenner, U., 1994. Ecological models of human development. *International encyclopedia of education*, 3(2), pp. 37–43.

—— 1977. Toward an experimental ecology of human development. *American psychologist*, 32(7), pp. 513–531.

Brusoni, S. & Prencipe, A., 2013. The organization of innovation in ecosystems: problem framing, problem solving, and patterns of coupling. *Advances in strategic management*, 30, pp. 167–194.

Ceccagnoli, M., Forman, C., Huang, P., Wu, D.J., 2012. Cocreation of value in a platform ecosystem: the case of enterprise software. *MIS quarterly*, 36(1). Available at: http://search.ebscohost.com/login.aspx?direct=true&profile=eho st&scope=site&authtype=crawler&jrnl=02767783&AN=71145739&h=ZflVtcm XT2f8279y9NRqM18HXs3D2Iuxe2SJnZJyb2xIw38r21Q2baPWfDC31OGZR5G 3m5SPRLzWMz%2FIxpc5Bg%3D%3D&crl=c [Accessed March 10, 2014].

Chang, V. & Uden, L., 2008. Governance for E-learning ecosystem. In *2nd IEEE International Conference on Digital Ecosystems and Technologies*, 2008. DEST 2008. 2nd IEEE International Conference on Digital Ecosystems and Technologies, 2008. DEST 2008. pp. 340–345.

Chen, C.J. & Chang, L.S., 2004. Dynamics of business network embeddedness. *Journal of American academy of business*, 5(1/2), pp. 237–241.

Chesbrough, H.W., 2003. *Open innovation: the new imperative for creating and profiting from technology*, Harvard Business Press.

Chesbrough, H., 2004. Managing open innovation. *Research-technology management*, 47(1), pp. 23–26.

Christopher, M., 1999. Logistics and supply chain management: strategies for reducing cost and improving service. *International journal of logistics research and applications*, 2(1), pp. 103–104.

—— 2005. *Logistics and supply chain management: creating value-added networks*, Pearson education, Great Britain.

Churchill, N.C. & Lewis, V., 1983. The five stages of business growth. *Harvard business review*, 61(3), pp. 30–50.

Darwin, C., 1859. *The origin of species*, Signet Classic, London.

Davis, T., 1993. Effective supply chain management. *Management review*, 8, pp. 35–46.

De Backer, K., 2008. *Open innovation in global networks*, Organisation for Economic Co-operation and Development, Paris.

Den Hartigh, E. & Van Asseldonk, T., 2004. Business ecosystems: a research framework for investigating the relation between network structure, firm strategy, and the pattern of innovation diffusion. In *ECCON 2004 Annual Meeting: Co-Jumping on a Trampoline*, The Netherlands.

Den Hartigh, E., Tol, M. & Visscher, W., 2006. The health measurement of a business ecosystem. In *Proceedings of the European network on chaos and complexity research and management practice meeting*, The Netherlands.

Desai, N., Mazzoleni, P. & Tai, S., 2007. Service communities: a structuring mechanism for service-oriented business ecosystems. In *Digital ecosystems and technologies conference, 2007*. DEST'07. Inaugural IEEE-IES, pp. 122–127.

Easterby-Smith, M., 1997. Disciplines of organizational learning: contributions and critiques. *Human relations*, 50(9), pp. 1085–1113.

Eisenhardt, K.M., 1989. Building theories from case study research. *Academy of management review*, 14(4), pp. 532–550.

Emerson, R.M., 1976. Social exchange theory. *Annual review of sociology*, 2(1), pp. 335–362.

Farhoomand, A.F., Ng, P.S. & Yue, J.K., 2001. The building of a new business ecosystem: sustaining national competitive advantage through electronic commerce. *Journal of organizational computing and electronic commerce*, 11(4), pp. 285–303.

Faulkner, D., 1995. *International strategic alliances: co-operating to compete*, McGraw-Hill, London.

Ferdows, K., 1997. Making the most of foreign factories. *Harvard business review*, 75, pp. 73–91.

—— 1989a. *Managing international manufacturing*, North Holland, Amsterdam.

—— 1989b. Mapping international factory networks. *Managing international manufacturing*, pp. 3–21.

Filson, D., 2001. The nature and effects of technological change over the industry life cycle. *Review of economic dynamics*, 4(2), pp. 460–494.

Finlay, P., 2002. *ARM: the chipless chip company, Case - Reference no. 302-026-1*, Available at: http://www.thecasecentre.org/educators/products/view?id=24907.

Fragidis, G., Koumpis, A. & Tarabanis, K., 2007. The impact of customer participation on business ecosystems. *Establishing the foundation of collaborative networks*, pp. 399–406.

Frosch, R.A. & Gallopoulos, N.E., 1989. Strategies for manufacturing. *Scientific American*, 261(3), pp. 144–152.

Garnsey, E., 1998. A theory of the early growth of the firm. *Industrial and corporate change*, 7(3), p. 523.

Garnsey, E., Lorenzoni, G. & Ferriani, S., 2008. Speciation through entrepreneurial spin-off: the Acorn-ARM story. *Research Policy*, 37(2), pp. 210–224.

Gassmann, O. & Enkel, E., 2004. Towards a theory of open innovation: three core process archetypes. In *R&D management conference*, pp. 1–18.

Gawer, A. & Cusumano, M.A., 2002. *Platform leadership: how Intel, Microsoft, and Cisco drive industry innovation*, Harvard Business School Pr, Boston.

—— 2013. Industry platforms and ecosystem innovation. *Journal of product innovation management*. Available at: http://onlinelibrary.wiley.com/doi/10.1111/jpim.12105/full [Accessed February 17, 2014].

Ghoshal, S. & Bartlett, C.A., 1990. The multinational corporation as an interorganizational network. *Academy of management review*, 15(4), pp. 603–625.

Gibbert, M., Ruigrok, W. & Wicki, B., 2008. What passes as a rigorous case study? *Strategic management journal*, 29(13), pp. 1465–1474.

Gassmann, O. & Enkel, E., 2004. Towards a theory of open innovation: three core process archetypes. In *R&D management conference*. University of St'Gallen, St' Gallen, Switzerland.

Greiner, L.E., 1997. Evolution and revolution as organizations grow. *Family business review*, 10(4), p. 397.

Gueguen, G., Pellegrin-Boucher, E. & Torres, O., 2006. Between cooperation and competition: the benefits of collective strategies within business ecosystems. The example of the software industry. In *EIASM, 2nd Workshop on competition strategy*, pp. 14–15.

Gulati, R., Puranam, P. & Tushman, M., 2012. Meta-organization design: rethinking design in interorganizational and community contexts. *Strategic management journal*, 33(6), pp. 571–586.

Hamel, G., 1991. Competition for competence and interpartner learning within international strategic alliances. *Strategic management journal*, 12(S1), pp. 83–103.

Hannan, M.T. & Freeman, J., 1977. The population ecology of organizations. *AJS*, 82(5), p. 929.

Harland, C., 1996. Supply network strategies: the case of health supplies. *European journal of purchasing & supply management*, 2(4), pp. 183–192.

Harvey, M.G., 1984. Application of technology life cycles to technology transfers. *Journal of business strategy*, 5(2), pp. 51–58.

Haupt, R., Kloyer, M. & Lange, M., 2007. Patent indicators for the technology life cycle development. *Research policy*, 36(3), pp. 387–398.

Hayes, R.H. & Wheelwright, S.C., 1984. *Restoring our competitive edge: competing through manufacturing*, John Wiley & Sons Inc., Hoboken.

—— 1979. The dynamics of process-product life cycles. *Harvard business review*, 57(2), pp. 127–136.

Hermens, A., 2003. Knowledge exchange in strategic alliances: learning in tension. *Creativity and innovation management*, 10(3), pp. 189–200.

Herriott, R.E. & Firestone, W.A., 1983. Multisite qualitative policy research: optimizing description and generalizability. *Educational researcher*, 12(2), p. 14.

Holm, D.B., Eriksson, K. & Johanson, J., 1999. Creating value through mutual commitment to business network relationships. *Strategic management journal*, 20(5), pp. 467–486.

Iansiti, M. & Levien, R., 2002. The new operational dynamics of business ecosystems: implications for policy, operations and technology strategy. *Harvard business school working paper*, Cambridge, MA, pp. 1–113.

—— 2004a. Strategy as ecology. *Harvard business review*, 82(3), pp. 68–81.

—— 2004b. *The keystone advantage: what the new dynamics of business ecosystems mean for strategy, innovation, and sustainability*. Harvard business school press, Cambridge, MA.

Iansiti, M. & Richards, G.L., 2006. The information technology ecosystem: structure, health, and performance. *Antitrust bulletin*, 51, p. 77.

Imielinski, T. & Korth, H.F., 1996. *Mobile computing*, Kluwer academic publishers, Norwell, MA.

Iyer, B., Lee, C.H. & Venkatraman, N., 2006. Managing in a small world ecosystem: some lessons from the software sector. *California management review*, 48(3), pp. 28–47.

Kapoor, R., 2013. Collaborating with complementors: What do firms do? *Advances in strategic management*, 30, pp. 3–25.

Kapoor, R. & Lee, J.M., 2013. Coordinating and competing in ecosystems: how organizational forms shape new technology investments. *Strategic management journal*, 34(3), pp. 274–296.

Kenney, M. & Pon, B., 2011. Structuring the smartphone industry: is the mobile internet OS platform the key? *Journal of industry, competition and trade*, 11(3), pp. 1–23.

Kim, B., 2003. Managing the transition of technology life cycle. *Technovation*, 23(5), pp. 371–381.

Klepper, S., 1996. Entry, exit, growth, and innovation over the product life cycle. *The American economic review*, 86(3), pp. 562–583.

—— 1997. Industry life cycles. *Industrial and corporate change*, 6(1), p. 145.

Klepper, S. & Miller, J.H., 1995. Entry, exit, and shakeouts in the United States in new manufactured products. *International journal of industrial organization*, 13(4), pp. 567–591.

Koza, M.P. & Lewin, A.Y., 1998. The co-evolution of strategic alliances. *Organization science*, 9(3), pp. 255–264.

Kraemer, K.L. & Dedrick, J., 2002. Strategic use of the internet and e-commerce: Cisco Systems. *Journal of strategic information systems*, 11(1), pp. 5–29.

Lambert, D.M. & Cooper, M.C., 2000. Issues in supply chain management. *Industrial marketing management*, 29(1), pp. 65–83.

Lawrence, E., 1990. *Henderson's dictionary of biological terms*, Longman, England.

Levitt, T., 1965. Exploit the product life cycle. *Harvard business review*, 43(6), pp. 81–94.

Li, J.F. & Garnsey, E., 2013. Building joint value: ecosystem support for global health innovations. *Advances in strategic management*, 30, pp. 69–96.

Li, Y.R., 2009. The technological roadmap of Cisco's business ecosystem. *Technovation*, 29(5), pp. 379–386

Li, X., Shi, Y. & Gregory, M., 2000. Global manufacturing virtual network (GMVN) and its position in the spectrum of strategic alliance. In *Operations management: crossing borders and boundaries: the changing role of operations*, EurOMA 7th International Annual Conference, Ghent, Belgium, pp. 330–337.

Lin, Y. & Zhou, L., 2011. The impacts of product design changes on supply chain risk: a case study. *International journal of physical distribution & logistics management*, 41(2), pp. 162–186.

Mäkinen, S.J. & Dedehayir, O., 2013. Business ecosystems' evolution–an ecosystem clockspeed perspective. *Advances in strategic management*, 30, pp. 99–125.

Manning, B., Runge, B., Thorne, C., Moore, G., 2002. *Demand driven: 6 steps to building an ecosystem of demand for your business*, McGraw-Hill Companies, New York.

Marín, C., Stalker, I. & Mehandjiev, N., 2007. Business ecosystem modelling: combining natural ecosystems and multi-agent systems. *Cooperative information agents*, 11, pp. 181–195.

Martin, F., 2010. *The new ICT ecosystem: implications for policy and regulation*, Cambridge University Press. Available at: http://dl.acm.org/citation.cfm?id=1855258 [Accessed December 4, 2012].

Messallam, A.A., 1998. The organizational ecology of investment firms in Egypt: organizational founding. *Organization studies*, 19(1), p. 23.

Miles, M.B. & Huberman, A.M., 1984. *Qualitative data analysis: A sourcebook of new methods*, Sage Publications, United States.

Mills, J., Platts, K. & Gregory, M., 1995. A framework for the design of manufacturing strategy processes. *International journal of operations & production management*, 15(4), pp. 17–49.

Minshall, T., Mortara, L., Elia, S., Probert, D., 2008. Development of practitioner guidelines for partnerships between start-ups and large firms. *Journal of manufacturing technology management*, 19(3), pp. 391–406.

—— 2010. Making 'asymmetric' partnerships work. *Research-technology management*, 53(3), pp. 53–63.

Mintzberg, H., Ahlstrand, B. & Lampel, J., 1998. *Strategy safari: a guided tour through the wilds of strategic management*, Free Press, New York.

Mitleton-Kelly, E., 2003. Ten principles of complexity and enabling infrastructures. *Complex systems and evolutionary perspectives on organisations: the application of complexity theory to organisations*, Elsevier, pp. 23–50.

Moore, J., 1993. Predators and prey: a new ecology of competition. *Harvard Business Review*, 71(3), pp. 75–86.

—— 1996. *The death of competition: leadership and strategy in the age of business ecosystems*, Harper Business, New York.

—— 2006. Business ecosystems and the view from the firm. *Antitrust Bulletin*, 51(1), p. 31.

Mortara, L., Slacik, I., Napp, J.J., Minshall, T., 2010. Implementing open innovation: cultural issues. *International Journal of Entrepreneurship and Innovation Management*, 11(4), pp. 369–397.

Mowery, D.C., Oxley, J.E. & Silverman, B.S., 1996. Strategic alliances and interfirm knowledge transfer. *Strategic management journal*, 17(1), pp. 77–91.

Nachira, F., 2002. Towards a network of digital business ecosystems fostering the local development. *Directorate General Information Society and Media, European Commission, Tech. Rep., 2002*. [Online]. Available: http://www.digital-ecosystems.org/doc/discussionpaper.pdf.

Nachira, F., Nicolai, A., Dini, P., Le Louarn, M., Leon, L.R., 2007. *Digital business ecosystems*. European Commission, Brussels. Retrieved 9 April 2008.

Nielsen, B.B., 2003. An empirical investigation of the drivers of international strategic alliance formation. *European management journal*, 21(3), pp. 301–322.

Niu, K.-H., 2009. The involvement of firms in industrial clusters: a conceptual analysis. *International Journal of Management*, 26(3), pp. 445–455.

Odum, E.P., 1971. *Fundamentals of ecology*, 3rd edn, Sounders, Philadelphia.

Oliver, R.K. & Webber, M.D., 1982. Supply-chain management: logistics catches up with strategy. *Logistics: the strategic issues*, Chapman & Hall, London, pp. 63–75.

Passiante, G. & Secundo, G., 2002. From geographical innovation clusters towards virtual innovation clusters: the Innovation Virtual System. In *ERSA conference*, University of Dortmund (Germany).

Peltoniemi, M., 2004. Cluster, value network and business ecosystem: knowledge and innovation approach. In 'Organisations, innovation and complexity: new perspectives on the knowledge economy' conference, September, University of Manchester, Manchester, England, UK, pp. 9–10.

—— 2005. Possibilities for research on the interaction of business models and success within a business ecosystem, Working paper. In Institute of Business Information Management, Tampere University of Technology, Finland.

—— 2006. Preliminary theoretical framework for the study of business ecosystems. *Emergence: complexity & organization*, 8(1), pp. 10–19.

—— 2011. Reviewing industry life-cycle theory: avenues for future research. *International journal of management reviews*, 13(4), pp. 349–375.

Peltoniemi, M. & Vuori, E., 2004. Business ecosystem as the new approach to complex adaptive business environments. *Frontiers of E-business research*, pp. 267–281. Tampere, Finland.

Peltoniemi, M., Vuori, E. & Laihonen, H., 2005. Business ecosystem as a tool for the conceptualisation of the external diversity of an organisation. In

Proceedings of the complexity, science and society conference, Liverpool, Great Britain, pp. 11–14.

Polli, R. & Cook, V., 1969. Validity of the product life cycle. *Journal of business*, 42(4), pp. 385–400.

Porter, M.E., 1979. How competitive forces shape strategy. *Harvard business review*, 57(2), pp. 137–145.

—— 1980. *Competitive strategy: techniques for analyzing industries and competitors*, Free Press, New York.

Porter, M.E., 1990. The competitive advantage of nations. *Harvard business review*, 68, pp. 73–93.

Porter, M.E., 1998. Clusters and the new economics of competition. *Harvard Business Review*, 76(6), pp. 77–90.

Porter, M.E. & Fuller, M.B., 1986. Coalitions and global strategy. *Competition in global industries*, 1(10), pp. 315–343.

Power, T. & Jerjian, G., 2001. *Ecosystem: living the 12 principles of networked business*, Financial Times/Prentice Hall, Harlow.

Quaadgras, A., 2005. Who joins the platform? The case of the RFID business ecosystem. In *System Sciences, 2005*. HICSS'05. Proceedings of the 38th Annual Hawaii International Conference on System Sciences, p. 269b.

Reid, S.E. & Brentani, U., 2012. Market vision and the front end of NPD for radical innovation: the impact of moderating effects. *Journal of Product Innovation Management*. Available at: http://onlinelibrary.wiley.com/doi/10.1111/j.1540-5885.2012.00955.x/full [Accessed October 9, 2012].

Rink, D.R. & Swan, J.E., 1979. Product life cycle research: a literature review. *Journal of business research*, 7(3), pp. 219–242.

Rogers, E.M., 1962. *Diffusion of innovations*, Free Press, New York.

Rohrbeck, R., Hölzle, K. & Gemünden, H.G., 2009. Opening up for competitive advantage: how Deutsche Telekom creates an open innovation ecosystem. *R&D management*, 39(4), pp. 420–430.

Rong, K., Shi, Y. & Zhang, Y., 2008. *CIKCCL industrial research: the semiconductor sector: exploring the essential elements of collaborative manufacturing networks*. Working paper.

Rong, K., Lin, Y., Shi, Y., Yu, J., 2013. Linking business ecosystem lifecycle with platform strategy: a triple view of technology, application and organization. *International journal of technology management*, 62(1), pp. 75–94.

Rong, K., Shi, Y. & Yu, J., 2013. Nurturing business ecosystems to deal with industry uncertainties. *Industrial management & data systems*, 133(3), pp. 385–402.

Rothschild, M., 1992. *Bionomics: economy as ecosystem*, Henry Holt & Company, New York.

Ruthertord, M.W., Buller, P.F. & McMullen, P.R., 2003. Human resource management problems over the life cycle of small to medium-sized firms. *Human resource management*, 42(4), pp. 321–335.

Schein, E.H., 2003. *DEC is dead, long live DEC: the lasting legacy of Digital Equipment Corporation*, Berrett-Koehler Publishers, San Francisco, California.

Scott, A.J., 1992. The role of large producers in industrial districts: a case study of high technology systems houses in Southern California. *Regional studies*, 26(3), pp. 265–275.

Shi, Y., 2004. A roadmap of manufacturing system evolution. In Y.S. Chang, H.C. Makatsoris, & H.D. Richards (eds), *Evolution of supply chain management*,

Springer US, New York, pp. 341–365. Available at: http://link.springer.com/chapter/10.1007/0-306-48696-2_14 [Accessed April 14, 2014].

Shi, Y. & Gregory, M., 1998. International manufacturing networks—to develop global competitive capabilities. *Journal of operations management,* 16(2–3), pp. 195–214.

—— 2001. Global manufacturing virtual network (GMVN): a new manufacturing system for market agility and global mobility. In International Working Conference on Strategic Manufacturing, 26–29 August. Aalborg, Denmark.

—— 2005. Emergence of global manufacturing virtual networks and establishment of new manufacturing infrastructure for faster innovation and firm growth. *Production planning & control,* 16(6), pp. 621–631.

Shi, Y.J., Fleet, D. & Gregory, M., 2003. Global manufacturing virtual network (GMVN): a revisiting of the concept after three years fieldwork. *Journal of Systems Science and Systems Engineering,* 12(4), pp. 432–448.

Shih, W. & Wang, J., 2009. *Upgrading the economy: industrial policy and Taiwan's semiconductor industry,* ecch case, Harvard business press, Boston.

Silberschatz, A., Galvin, P.B., Gagne, G., 1991. *Operating system concepts,* Addison-Wesley, New York.

Simonin, B.L., 1999. Ambiguity and the process of knowledge transfer in strategic alliances. *Strategic management journal,* 20(7), pp. 595–623.

Srai, J.S. & Gregory, M., 2008. A supply network configuration perspective on international supply chain development. *International journal of operations & production management,* 28(5), pp. 386–411.

Thomas, D.J. & Griffin, P.M., 1996. Coordinated supply chain management. *European journal of operational research,* 94(1), pp. 1–15.

Van der Vorst, J.G.A.J. & Beulens, A.J.M., 2002. Identifying sources of uncertainty to generate supply chain redesign strategies. *International journal of physical distribution and logistics management,* 32(6), pp. 409–430.

Van der Vorst, J.G.A.J., Beulens, A.J.M., Wit, W., Beek, P., 2006. Supply chain management in food chains: Improving performance by reducing uncertainty. *International transactions in operational research,* 5(6), pp. 487–499.

Van Witteloostuijn, A., 2000. Organizational ecology has a bright future. *Organization studies,* 21(2), pp. 5–14.

Vernon, R., 1966. International investment and international trade in the product cycle. *The quarterly journal of economics,* 80(2), pp. 190–207.

Von Bertalanffy, L., 1950. An outline of general system theory. *The British journal for the philosophy of science,* 1(2), p. 134.

—— 1969. *General system theory: foundations, development, applications,* G. Braziller, New York.

Wernerfelt, B., 1984. A resource-based view of the firm. *Strategic management journal,* 5(2), pp. 171–180.

West, J. & Wood, D., 2013. Evolving an open ecosystem: the rise and fall of the Symbian platform. *Advances in strategic management,* 30, pp. 27–67.

Wilding, R., 1998. The supply chain complexity triangle: uncertainty generation in the supply chain. *International journal of physical distribution and logistics management,* 28, pp. 599–616.

Wilkinson, I. & Young, L., 2002. On cooperating: firms, relations and networks. *Journal of Business Research,* 55(2), pp.123–132.

Williamson, O.E., 1975. *Markets and hierarchies: analysis and antitrust implications*, Free Press, New York.

—— 1995. Transaction cost economics and organization theory. *Organization theory: from Chester Barnard to the present and beyond*, New York, Oxford University Press, pp. 207–256.

Williamson, P.J. & De Meyer, A., 2012. Ecosystem advantage: how to successfully harness the power of partners. *California management review*, 55(1), pp. 24–46.

Wood, D., Northam, P. & Gjertsen, F., 2005. *Symbian for software leaders*, Wiley, Hoboken.

Yin, R., 2008. *Case study research: Design and methods*, Sage Publications, United States.

Zhang, Y., Gregory, M. & Shi, Y.J., 2007. Global engineering networks: the integrating framework and key patterns. *Proceedings of the institution of mechanical engineers, part B: journal of engineering manufacture*, 221(8), pp. 1269–1283.

Zhu, S. & Shi, Y., 2010. Shanzhai manufacturing–an alternative innovation phenomenon in China: its value chain and implications for Chinese science and technology policies. *Journal of science and technology policy in China*, 1(1), pp. 29–49.

Index

CPI Antony Rowe
Eastbourne, UK
June 18, 2020